My
Windows® 11
Computer
for Seniors

 |

Michael Miller

My Windows® 11 Computer for Seniors

Copyright © 2022 by Pearson Education, Inc.

ISBN-13: 978-0-13-784170-7

ISBN-10: 0-13-784170-1

Library of Congress Control Number: 2022930140

1 2022

Trademarks

Warning and Disclaimer

Special Sales

For information about buying this title in bulk quantities, or for special sales opportunities (which may include electronic versions; custom cover designs; and content particular to your business, training goals, marketing focus, or branding interests), please contact our corporate sales department at corpsales@pearsoned.com or (800) 382-3419.

For government sales inquiries, please contact governmentsales@pearsoned.com.

For questions about sales outside the U.S., please contact intlcs@pearson.com.

Editor-in-Chief
Brett Bartow

Executive Editor
Laura Norman

Director, AARP Books
Jodi Lipson

Associate Editor
Anshul Sharma

Editorial Services
The Wordsmithery LLC

Managing Editor
Sandra Schroeder

Senior Project Editor
Tonya Simpson

Copy Editor
Charlotte Kughen

Indexer
Cheryl Lenser

Proofreader
Sarah Kearns

Technical Editor
Vince Averello

Editorial Assistant
Cindy J. Teeters

Designer
Chuti Prasertsith

Compositor
Bronkella Publishing

Graphics
TJ Graham Art

Contents at a Glance

Table of Contents

Dedication

To Sherry. As always.

About the Author

Michael Miller is a prolific and popular writer of more than 200 nonfiction books who is known for his ability to explain complex topics to everyday readers. He writes about a variety of topics, including technology, business, and music. His best-selling books for Que and AARP include *My Video Chat for Seniors, My iPad for Seniors, My TV for Seniors, My Social Media for Seniors, My Facebook for Seniors, My Smart Home for Seniors, My Internet for Seniors,* and *My eBay for Seniors.* Worldwide, his books have sold more than 1.5 million copies.

Find out more at the author's website: www.millerwriter.com

Follow the author on Twitter: @molehillgroup

Acknowledgments

Thanks to all the folks at Que and Pearson who helped turn this manuscript into a book, including Laura Norman, Anshul Sharma, Charlotte Kughen, Tonya Simpson, and technical editor Vince Averello. Thanks also to Jodi Lipson and the good folks at AARP for supporting and promoting this and other books.

Pearson's Commitment to Diversity, Equity, and Inclusion

Pearson is dedicated to creating bias-free content that reflects the diversity of all readers. We embrace the many dimensions of diversity, including but not limited to race, ethnicity, gender, socioeconomic status, ability, age, sexual orientation, and religious or political beliefs.

Books are a powerful force for equity and change in our world. They have the potential to deliver opportunities that improve lives and enable economic mobility. As we work with authors to create content for every product and service, we acknowledge our responsibility to demonstrate inclusivity and incorporate diverse scholarship so that everyone can achieve their potential through learning. As the world's leading learning company, we have a duty to help drive change and live up to our purpose to help more people create a better life for themselves and to create a better world.

Our ambition is to purposefully contribute to a world where:

- Everyone has an equitable and lifelong opportunity to succeed through learning.
- Our products and services are inclusive and represent the rich diversity of readers.
- Our content accurately reflects the histories and experiences of the readers we serve.
- Our content prompts deeper discussions with readers and motivates them to expand their own learning (and worldview).

While we work hard to present unbiased content, we want to hear from you about any concerns or needs with this Pearson product so that we can investigate and address them. Please contact us with concerns about any potential bias at https://www.pearson.com/report-bias.html.

About AARP

AARP is the nation's largest nonprofit, nonpartisan organization dedicated to empowering people 50 and older to choose how they live as they age. With a nationwide presence and nearly 38 million members, AARP strengthens communities and advocates for what matters most to families: health security, financial stability and personal fulfillment. AARP also produces the nation's largest circulation publications: *AARP The Magazine* and *AARP Bulletin*. To learn more, visit www.aarp.org, www.aarp.org/espanol or follow @AARP, @AARPenEspanol and @AARPadvocates, @AliadosAdelante on social media.

NOTE

Most of the individuals pictured throughout this book are the author himself, as well as friends and relatives (used with permission) and sometimes pets. Some names and personal information are fictitious.

We Want to Hear from You!

As the reader of this book, *you* are our most important critic and commentator. We value your opinion and want to know what we're doing right, what we could do better, what areas you'd like to see us publish in, and any other words of wisdom you're willing to pass our way.

You can email to let us know what you did or didn't like about this book—as well as what we can do to make our books better.

Please note that we cannot help you with technical problems related to the topic of this book.

When you write, please be sure to include this book's title and author, as well as your name, email address, and phone number. We will carefully review your comments and share them with the author and editors who worked on the book.

Email: community@informit.com

Reader Services

Register your copy of *My Windows 11 Computer for Seniors* at informit.com/register for convenient access to downloads, updates, and corrections as they become available. To start the registration process, go to informit.com/register and log in or create an account.* Enter the product ISBN (9780137841707) and click Submit.

*Be sure to check the box that you would like to hear from us to receive exclusive discounts on future editions of this product.

Figure Credits

Cover art INGARA/Shutterstock

Chapter 1 opener image of couple: gpointstudio/123RF

Chapter 5, screenshots from TCL © 2022 TCL

Chapter 8, pull-down menu image © 2022 Intuit Inc.

Chapter 11, screenshots from Google © 2022 Google LLC

Chapter 11, screenshots from DuckDuckGo © 2022 DuckDuckGo

Chapter 12, Sephora web page © Sephora

Chapter 12, Best Buy web page © Best Buy

Chapter 12, Odeals web page © Odeals

Chapter 12, Kohl's web page © Kohl's

Chapter 12, screenshots from Lands' End web page © Lands' End

Chapter 12, image from Craigslist © Craigslist

Chapter 12, eBay web page © eBay

Chapter 12, Etsy web page © Etsy

Chapter 12, Facebook Marketplace web page © Facebook

Chapter 12, Reverb web page © Reverb

Chapter 12, DoorDash web page © DoorDash

Chapter 12, Club web page © Club

Chapter 13, screenshots from Facebook © 2022 Meta

Chapter 14, screenshots from Gmail © 2022 Google LLC

Chapter 15, screenshots from Zoom Video Communications © 2022 Zoom Video Communications, Inc.

Chapter 17, screenshots from Facebook © 2022 Meta

Chapter 17, screenshots from Pinterest © 2022 Pinterest

Chapter 17, screenshots from Twitter © 2022 Twitter, Inc.

→ Examining Key Components
→ Exploring Different Types of PCs
→ Which Type of PC Should You Buy?
→ Setting Up Your New Computer System

Understanding Computer Basics

What should you look for if you need a new computer? What are all those pieces and parts? And how do you connect everything?

These are common questions for anyone just getting started with a personal computer (PC)—whether a desktop, laptop (portable), all-in-one, or 2-in-1 (which I explain later in this chapter). Read on to learn more about the key components of a typical computer system—and how they all work together.

Examining Key Components

All computers do pretty much the same things, in pretty much the same ways. There are differences, however, in the capacities and capabilities of key components, which can affect how fast your computer operates. And when you're shopping for a new PC, you need to keep these options in mind.

Hard Disk Drive

All computers feature some form of long-term storage for your documents, photos, music, and videos. On many desktop PCs and some laptop computers, this storage is in the form of an *internal hard disk drive*. This is a device that stores data magnetically, on multiple metallic platters—kind of like a high-tech electronic juke box.

Most computers today come with large hard drives, ranging in size from 256 gigabytes (GB) to 1 terabyte (TB) or more. The more storage space the better, especially if you have lots of digital photos or videos to store.

Kilobytes, Megabytes, Gigabytes, and Terabytes

The most basic unit of digital storage is called a byte; a byte typically equals one character of text. One thousand bytes equal one *kilobyte* (KB). One thousand kilobytes, or one million bytes, equal one *megabyte* (MB). One thousand megabytes, or one billion bytes, equal one *gigabyte* (GB). One thousand gigabytes, or one trillion bytes, equal one *terabyte* (TB).

Solid-State Drive

These days, hard disk drives are being supplanted by a different type of long-term storage that uses solid-state technology. A solid-state drive (SSD) has no moving parts; instead, it stores data electronically on an integrated circuit. This type of storage is both lighter and faster than traditional hard disk storage; data stored on an SSD can be accessed pretty much instantly. Plus, computers with SSDs weigh considerably less than models with traditional hard drives, which is especially important for laptop and 2-in-1 models.

Lettered Drives

All storage drives, whether hard disk or solid state drive (SSD), are identified by letters assigned by Windows. On most systems, the main drive is called the c: drive. If you have a second drive, it will be the d: drive. Any external drives attached to your PC will pick up the lettering from there.

The downside of solid-state storage is that it's a little more expensive than hard drive storage. What this means is that you typically get a little less storage on an SSD than you would on a similar computer with a traditional hard drive—or you pay more for a computer with similarly sized SSD.

SSD and Hard Drive Storage

Some computers offer both SSD and hard drive storage. The smaller SSD is typically used to house the Windows operating system for super-fast start-up and speedy operation. Data is stored longer term on the larger hard disk drive.

Memory

Hard disks and solid-state memory devices provide long-term storage for your data. Your computer also needs short-term storage to temporarily store documents as you're working on them or photos you're viewing.

This short-term storage is provided by your PC's *random access memory*, or RAM. Most PCs today offer anywhere from 4 to 32 gigabytes (GB) of RAM. The more memory in your computer, the faster it operates.

System Unit

On a traditional desktop computer, the hard disk, memory, and CPU are contained within a separate *system unit* that also sports various connectors and ports for monitors and other devices. On an all-in-one desktop, the system unit is built in to the monitor display. On a laptop or 2-in-1 PC, the hard disk and other components are part of the laptop itself.

Processor

The other major factor that affects the speed of your PC is its *central processing unit* (CPU) or *processor*. The more powerful your computer's CPU, measured in terms of gigahertz (GHz), the faster your system runs.

Today's CPUs often contain more than one processing unit. A dual-core CPU contains the equivalent of two processors in one unit and should be roughly twice as fast as a comparable single-core CPU; a quad-core CPU should be four times as fast as a single-core CPU.

>>>Go Further
TPM FOR WINDOWS 11

In addition to the standard CPU, most newer computers include a special cryptoprocessor chip called Trusted Platform Module (TPM). The correct version of this chip is necessary to run Windows 11. The TPM chip uses an integrated cryptographic key to secure your computer hardware. That sounds complicated (and it is, a little), but it basically works like a security alarm that keeps malicious software (malware) and hackers from accessing your computer's data.

Computers older than 15 years or so don't have TPM chips, and Windows 11 requires a TPM chip to run. If your computer doesn't have a TPM chip, you can't upgrade it to Windows 11; it must keep running the older Windows 10 operating system.

Complicating things even more, Windows 11 requires the newest version of TPM to run. TPM 2.0 was introduced in 2014, and if you have an older computer that isn't running that version, it can't be upgraded to Windows 11.

Unfortunately, TPM isn't something you can upgrade. Your computer either has TPM 2.0 (and can run Windows 11) or it doesn't (and it can't). To check if your older computer has TPM 2.0 and is capable of running Windows 11, download and launch Microsoft's PC Health Check app from the main Windows 11 web page, located at www.microsoft.com/en-us/windows/windows-11.

Display

All computers today come with liquid crystal display (LCD) screens. The screen can be in an external monitor in desktop systems, combined with the system unit for all-in-one systems, or built into a laptop or 2-in-1 PC. Screens come in a variety of sizes, from 10" diagonal in small laptop PCs to 34" diagonal or more in larger desktop systems. There are even ultrawide monitors with 49" diagonal screens! Naturally, you should choose a screen size that fits on your desk and makes it easy for you to read and view videos.

Some LCD monitors, especially those on laptop and 2-in-1 PCs, offer touch-screen operation. With a touchscreen, you can perform many operations with the tap or swipe of a fingertip. Because touchscreen displays cost more than traditional displays, they're typically not on lower-end models.

External Monitor for a Laptop PC

Most users are happy with the built-in display in their laptops and 2-in-1 PCs. If you prefer a larger display, however, it's easy to connect an external LCD monitor to your laptop, via the laptop's HDMI port. (Learn more about HDMI in the "Connectors" section later in this chapter.)

Touchpads for Touchscreens

Some touchpads on laptop and 2-in-1 PCs let you emulate a touchscreen display. That is, you can perform similar touch gestures on one of these touchpads as you can on a touchscreen. (Learn more about touchpad input later in this chapter in the "Pointing Device" section.)

Keyboard

When it comes to typing letters, emails, and other documents, as well as post-ing updates to websites such as Facebook, you need an alphanumeric keyboard. On a desktop or all-in-one PC, the keyboard is an external component (called a *peripheral*); the keyboard is built in to all laptop and 2-in-1 PCs.

Function keys
External keyboard
Numeric keypad
Windows key
Arrow (direction) keys

Computer keyboards include typical typewriter keys, as well as a set of so-called *function keys* (designated F1 through F12) aligned on the top row of the keyboard; these function keys provide one-touch access to many computer functions. For example, pressing the F1 key in many programs brings up the program's help system.

Also, several keys that aren't letters or numbers are used to perform general functions. For example, the Escape (Esc) key typically undoes the current action, the Backspace key deletes the previous character, and the Delete (Del) key deletes the current character. And, as I explain later in this book, there are also Windows and Menu keys that have specific functionality within the Windows operating system.

In addition, most external (and some laptop) keyboards have a separate numeric keypad, which makes it easier to enter numbers. There are also number keys beneath the function keys on all computer keyboards.

External Input on a Laptop PC

Even though laptop and 2-in-1 PCs come with built-in keyboards and touchpads, you can still connect external keyboards and mice (pointing devices) if you like, via the PC's USB ports or wirelessly via Bluetooth technology. (Read more about USB ports later in this chapter in the "Connectors" section.) Some users prefer the feel of a full-size keyboard and mouse to the smaller versions included in their laptops.

Pointing Device

You use a pointing device of some sort to move the cursor from place to place on the computer screen. On a desktop PC, the pointing device of choice is called a *mouse*; it's about the size of a bar of soap, and you make it work by rolling it across a hard surface, such as a desktop.

Scroll wheel

Left button

Right button

External mouse

Most laptop PCs have a built-in pointing device called a *touchpad*. You move your fingers across the touchpad to move the cursor across the computer screen.

Both mice and touchpads have accompanying buttons that you click to perform various operations. Some devices include both left and right buttons; clicking the left button activates most common functions, whereas clicking the right button provides additional functionality in select situations.

Touchpad

Right-click area

Left-click area

Many touchpads don't have discrete buttons. Instead, the lower part of the touchpad is designated as the button area; you tap on the lower-left quadrant to left-click and tap on the lower-right quadrant to right-click.

Connectors

Every computer comes with a variety of connectors (called *ports*) to which you can connect external components (called *peripherals*), such as keyboards, printers, and the like. A number of different connectors are available, and not all computers offer the same assortment.

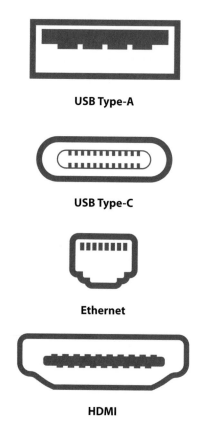

USB Type-A

USB Type-C

Ethernet

HDMI

On today's computers, the most common type of connector is called the *universal serial bus*, or USB. Most external devices connect to your computer via USB.

There are two types of USB ports you might find on your PC. The most common is the larger USB Type-A connector that's used to connect printers and other traditional peripherals. Some newer PCs also include the smaller USB Type-C connector, also found on many smartphones and tablets, which is used to connect small peripherals.

Your computer also has one or more connectors for an external monitor. (Most laptops also have one of these connectors, even though they have built-in monitors.) In most new PCs, this is a *high-definition multimedia interface* (HDMI) connector, like the ones you have on your living room TV. Because HDMI transmits both video and audio, you can use this port to connect your computer to your living room TV. (HDMI is also used to connect Blu-ray players, cable boxes, and other devices to television sets.)

Finally, many computers have an *Ethernet* port to connect to wired home and office networks. Most PCs today also offer wireless network connectivity, via a technology called *Wi-Fi*. If your computer has Wi-Fi (and you have Wi-Fi at your location), you don't have to connect via a cable.

Exploring Different Types of PCs

If you're in the market for a new PC, you'll find four general types available—traditional desktops, all-in-ones, traditional laptops, and 2-in-1s. All types of computers do pretty much the same thing, and they do it in similar ways; the differences between desktop and laptop computers are more about how they're configured than how they perform.

Desktop PCs

The first general type of PC is the *traditional* desktop system. A desktop computer is designed to be used in one place. It's a stationary computer rather than a portable one.

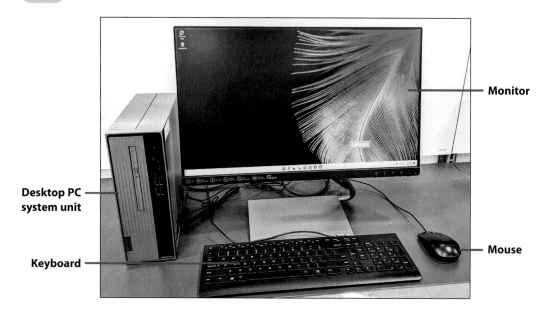

All desktop PCs have a separate keyboard and mouse to use for typing and navigating the screen. You also have a monitor, or computer screen, and a system unit that houses all the internal electronics for the entire system. You can store the system unit under your desk or in some other out-of-the-way place. (If you do place the system unit under your desk, make sure it has sufficient airflow to avoid overheating.)

For many users, the major advantage of a traditional desktop is the larger monitor screen, full-size keyboard, and separate mouse. It's easier to read many documents on a larger desktop monitor, and most full-size keyboards also offer numeric keypads, which are easier for entering numbers when you're doing online banking or budgeting. Many users find the separate mouse easier to use than the small touchpad found on most laptop PCs.

Monitor Screens

Most desktop computer monitors have screens that measure 19" to 24" diagonally. Most laptop PC screens measure 10" to 16" diagonally, so they're considerably smaller than their desktop counterparts. However, an external large screen monitor can easily be connected to a laptop PC.

On the downside, a desktop PC isn't portable; you have to leave it in one place in your home or office. In addition, a desktop system is a little more complicated to set up, with all its external components. What's more, you'll likely pay a little more for a desktop system than you will for a similarly configured laptop PC.

All-in-One PCs

Traditional desktop PCs have pretty much been supplanted by *all-in-one* desktops. An all-in-one PC builds the system unit and speakers into the monitor for a more compact, space-saving system. Some all-in-one PCs feature touchscreen monitors, so you can control them by tapping and swiping the monitor screen.

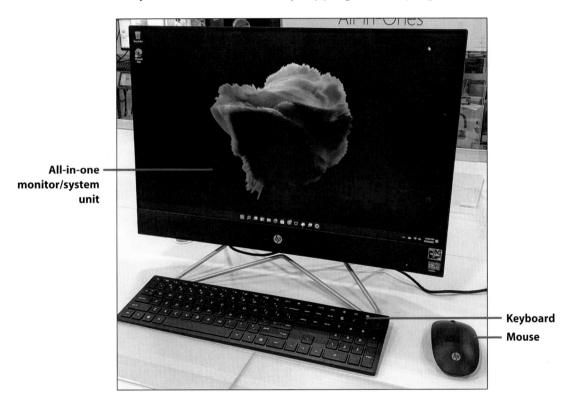

All-in-one monitor/system unit

Keyboard
Mouse

A lot of folks like the easier setup (no system unit or speakers to connect) and smaller space requirements of all-in-one systems. The drawbacks to these all-in-one desktops are that you can't upgrade internal components, and if one component (such as the screen) goes bad, the entire system is out of commission. It's a lot easier to replace a single component in a traditional desktop than the entire system of an all-in-one.

Laptop PCs

A *laptop* PC, sometimes called a notebook computer, combines all the components of a desktop system into a single unit with built-in screen, keyboard, and touchpad. Laptop PCs are not only small and lightweight, often less than four pounds, but also portable because they're capable of operating from a built-in battery that can last anywhere from 2 to 6 hours on a charge. (Naturally, a laptop PC can also be plugged in to a wall to use standard AC power.)

In addition, you can take a laptop PC just about anywhere. You can move your laptop to your living room or bedroom as you desire, and even take it with you when you're traveling or use it in public places such as coffeehouses or airports.

On the downside, the typical laptop PC has a smaller screen than a desktop system, which can make it more difficult to view smaller items onscreen. In addition, the compact keyboard of a laptop model might make typing more difficult for some people. Most laptop PCs also use a small touchpad to navigate onscreen, as opposed to the larger mouse of a desktop system, which some people might find difficult to use. (You can always connect an external mouse to your laptop, as discussed later in this chapter.)

2-in-1 PCs

Many laptop computers today combine the features of a traditional laptop with those of a tablet. (*Tablets* are portable touchscreen devices, such as the Amazon Fire or Apple iPad devices.) These *2-in-1 PCs*, as they're called, typically let you swivel the display against the keyboard to emulate touchscreen tablet operation or swivel the display the other way to let you use the device with the traditional keyboard.

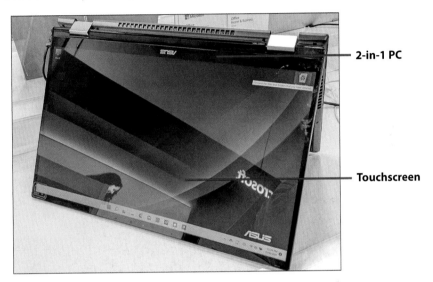

2-in-1 PC

Touchscreen

A 2-in-1 is great if you need the functionality of a laptop PC and the portability of a tablet—even if they do cost a bit more than a normal laptop. Many users find the touchscreen operation particularly nice.

Which Type of PC Should You Buy?

Which type of PC you purchase depends on how and where you plan to use your new computer. Here are some recommendations:

- If you need a larger screen, prefer a full-sized keyboard and mouse, and don't need your computer to be portable, go with a desktop PC. Consider an all-in-one system for easier setup.

- If you don't want to bother with connecting cables and external devices, go with an all-in-one, laptop, or 2-in-1 PC.

- If you want to take your computer with you when you move from room to room, go to the coffeeshop, or travel, go with a laptop or 2-in-1 PC.

- If you want to use your PC as a tablet (with a touchscreen and no keyboard), go with a 2-in-1 PC.

Naturally, every person has his or her unique needs and preferences. Always try out a system in the store to see if it's comfortable for you before making a purchase.

Setting Up Your New Computer System

After you purchase a new PC, you need to set up and connect the system's hardware. As you might suspect, this is easier to do with a laptop PC than it is with a desktop system.

Hardware and Software

All the physical parts of your computer—the screen, the system unit, the keyboard, and so forth—are referred to as *hardware*. The programs, apps, and games you run on your computer are called *software*.

Set Up a Laptop or 2-in-1 PC

If you have a laptop PC, there isn't much you need to connect; everything's inside the case. Just connect your printer (and any other external peripherals, such as a mouse if you prefer to use one instead of a touchpad) either wirelessly or physically via USB, plug your laptop into a power outlet, and you're ready to go.

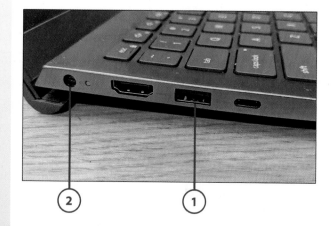

1. If you have any peripherals to connect, such as a printer, do so first. (Most peripherals connect via USB.)

2. Connect one end of your computer's power cable to the power connector on the side or back of your laptop. (You can skip this and the next step after initial setup if you're running on battery power.)

3 Connect the other end of the computer's power cable to a power source and then connect any powered external peripherals to the same power source.

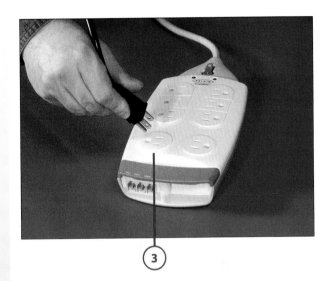

Use a Surge Suppressor

For extra protection, don't plug the power cable on your laptop or desktop (or any powered external peripherals, such as a monitor and printer) directly into an electrical outlet. Instead, connect the power cable to a power strip that incorporates a surge suppressor. This protects your PC from power-line surges that can damage its delicate internal parts.

Set Up an All-in-One PC

Setting up an all-in-one desktop PC is a little like setting up a laptop PC because the speakers, monitor, and system unit are in a single unit. The only extra things you have to connect are the mouse, keyboard, and any external peripherals you might have, such as a printer. This makes for a relatively quick and easy setup.

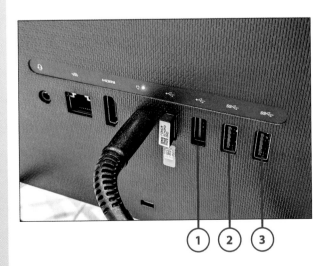

1 Connect the mouse cable to a USB port on the monitor.

2 Connect the keyboard cable to a USB port on the monitor.

3 Connect one end of your printer's USB cable to a USB port on the monitor; connect the other end of the cable to your printer. (If your printer connects wirelessly via Wi-Fi, skip this step.)

(4) Connect one end of your computer's power cable to the power connector on the monitor.

(5) Connect the other end of the power cable to a power source and then connect your printer and other powered external peripherals to the same power source.

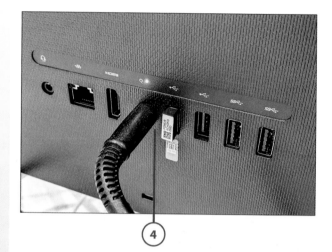

4

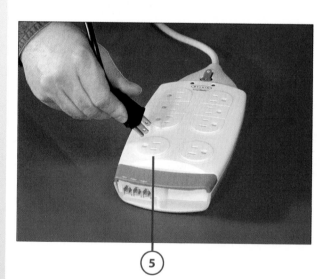

5

Set Up a Traditional Desktop PC

If you have a traditional desktop computer system, you need to connect all the pieces and parts to your computer's system unit before powering it on. When all your peripherals are connected, you can connect your system unit to a power source.

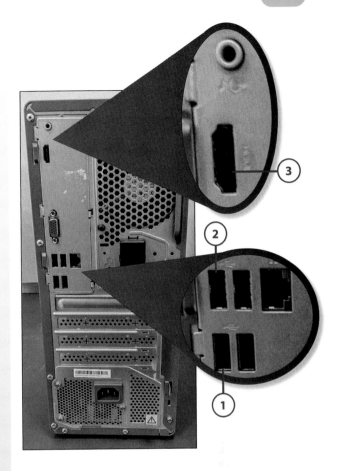

① Connect the mouse cable to a USB port on the back of your system unit.

② Connect the keyboard cable to a USB port on the back of your system unit.

③ Connect an HDMI cable to the corresponding port on the back of your system unit. Make sure the other end is connected to your video monitor.

Older Monitors

If you have an older monitor without an HDMI output, you can connect it to your PC's VGA input instead. (VGA is an older cable standard used to connect computer monitors; newer monitors use HDMI instead.)

(4) Connect the green phono cable from your main external speaker to the audio-out or sound-out connector on your system unit; connect the other end of the cable to the speaker. (Some external speakers connect via USB, which is even simpler; just connect the speaker cable to an open USB port on your system unit.)

(5) Connect one end of your printer's USB cable to a USB port on the back of your system unit; connect the other end of the cable to your printer. (If your printer connects wirelessly via Wi-Fi, skip this step.)

(6) Connect one end of your computer's power cable to the power connector on the back of your system unit.

(7) Connect the other end of the power cable to a power source and then connect your printer, monitor, and any other powered external peripherals to the same power source.

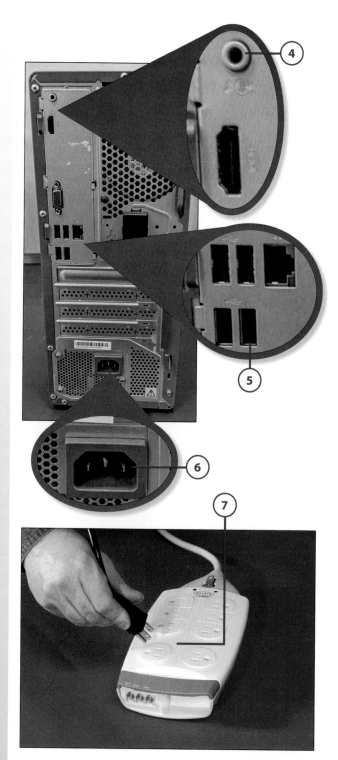

In this chapter, you find out how to operate Windows with your mouse, touchpad, keyboard, or touchscreen.

→ Using Windows with a Mouse or Touchpad
→ Using Windows with a Keyboard
→ Using Windows with a Touchscreen Display

2

Performing Basic Operations

Whether you're completely new to computers or just new to Windows 11, you need to master some basic mouse, keyboard, and (if you have a touchscreen display) touch operations to use your PC.

Using Windows with a Mouse or Touchpad

To use Windows efficiently on a desktop or laptop PC, you need to master a few simple operations with your mouse or touchpad, such as pointing and clicking, dragging and dropping, and right-clicking.

Mouse and Touchpad Operations

Of the various mouse and touchpad operations, the most common is pointing and
clicking—that is, you point at something
with the onscreen cursor and then click or
tap the appropriate mouse or touchpad
button. Normal clicking or tapping uses
the left mouse button or lower-left portion
of the touchpad; however, some operations require that you click or tap the right
button or lower-right area of the touchpad
instead.

(1) To single-click (select) an item, position the cursor over the onscreen item
and click or tap the left mouse button
or lower-left area of the touchpad.

(2) To double-click (select or open) an
item, position the cursor over the
onscreen item and click or tap the left
mouse button or lower-left area of the
touchpad twice in rapid succession.

(3) To right-click an item (to display a
context-sensitive options menu), position the cursor over the onscreen item
and then click or tap the *right* mouse
button or lower-right area of the
touchpad.

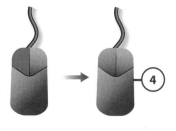

(4) To drag and drop an item from one
location to another, position the cursor over the item, click or tap and hold
the left mouse button or lower-left
area of the touchpad, drag the item to
a new position, and then release the
mouse button or lift your finger from
the touchpad.

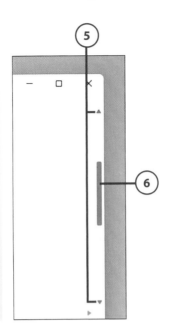

5 To scroll through a window, use your mouse or touchpad to hover over the window and display the scroll bar; then move the cursor over the up or down arrow on the scroll bar and click or tap the left mouse button or lower-left area of the touchpad.

6 To move to a specific place in a long window or document, click the scroll box (between the up and down arrows) and drag it to a new position.

Scroll Wheel

If your mouse has a scroll wheel, you can use it to scroll through a long document. Just roll the wheel back or forward to scroll down or up through a window. Likewise, some laptop touchpads let you drag your finger up or down to scroll through a window.

Mouse Over

Another common mouse operation is called the *mouse over*, or *hovering*, where you hold the cursor over an onscreen item without pressing either of the mouse buttons or tapping the touchpad. For example, when you mouse over an icon or menu item, Windows displays a *ToolTip* that tells you a little about the selected item.

Using Windows with a Keyboard

You can perform many operations in Windows without using your mouse or touchpad. Many users prefer to use their keyboards because it lets them keep their hands in one place when they're entering text and other information.

Keyboard Operations

Many Windows operations can be achieved from your computer keyboard, without touching your mouse or touchpad. Several of these operations use special keys that are unique to Windows PC keyboards, such as the Windows and Application keys.

 To scroll down any page or screen, press the PageDown key.

 To scroll up any page or screen, press the PageUp key.

   To launch a program or open a file, use the keyboard's arrow keys to move to the appropriate item and then press the Enter key.

 To display a context-sensitive pop-up menu (the equivalent of right-clicking an item), use the keyboard's arrow keys to move to that item and then press the Application key.

 To cancel or "back out" of the current operation, press the Escape key.

 To rename a file, use the keyboard's arrow keys to move to that file and then press the F2 key.

 To access an application's Help system, press the F1 key.

 To display the Start menu, press the Windows key.

Using Windows with a Touchscreen Display

If you're using Windows 11 on a computer with a touchscreen display, you can use your fingers on the screen to do what you need to do. To that end, it's important to learn some essential touchscreen operations.

Touchscreen Operations

Many operations in Windows 11 can be performed without a mouse or keyboard, using simple touch gestures instead.

 To "click" or select an item on a touchscreen display, tap the item with the tip of your finger and release.

 To "right-click" an item on a touchscreen display (typically displays a context-sensitive options menu), press, hold, and then release the item with the tip of your finger.

 To scroll up or down a page, swipe the screen in the desired direction.

 To zoom in on a given screen (that is, to make a selection larger), use two fingers to touch two points on the item, and then move your fingers apart.

 To zoom out of a given screen (that is, to make a selection smaller and see more of the surrounding page), use two fingers (or your thumb and first finger) to touch two points on the item, and then pinch your fingers in toward each other.

>>>Go Further
2-IN-1 COMPUTERS AND TABLET OPERATION

As you learned in Chapter 1, "Understanding Computer Basics," some computers have a display that folds back so you can use the device as either a computer with a keyboard or a touch-screen tablet. These 2-in-1 PCs are quite versatile and offer a similar experience whichever way you're using them—with some subtle differences.

When you're using a 2-in-1 PC in Tablet mode, the Windows 11 desktop adjusts slightly to make it easier to navigate with just your fingers. This so-called *tablet experience* displays the normal Windows desktop, including a traditional taskbar, but with a little more spacing between onscreen elements.

In Windows 10, Microsoft offered a distinct Tablet mode for 2-in-1 and tablet users. That mode offered an entirely different desktop layout, complete with big tiles in place of app icons. Microsoft nixed Tablet mode in Windows 11 in favor of the slightly different automatic layout changes because the company wanted to offer similar experiences for both desktop and tablet operation.

Desktop

Open app window

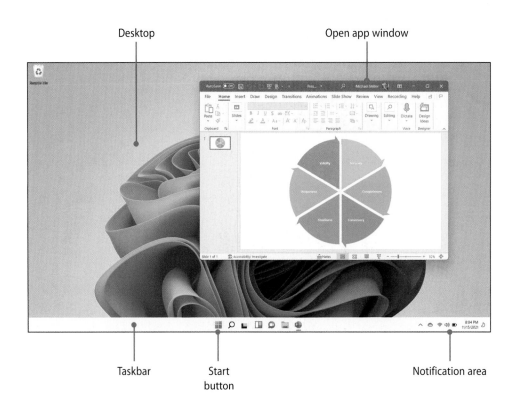

Taskbar

Start
button

Notification area

In this chapter, you find out how to turn on and start using a new Windows 11 computer.

→ Powering Up and Powering Down
→ Finding Your Way Around Windows
→ Switching from Windows S Mode to Windows Home

Using Your Windows 11 PC

Whether you've been using computers forever or just purchased your first PC, there's a lot you need to know about using the Windows operating system—such as where everything is, what it does, and how to do what you need to do.

Powering Up and Powering Down

If you've already read Chapter 1, "Understanding Computer Basics," you know how to connect all the components of your new computer system. The first time you turn on a new computer, you're led through a series of steps to configure the computer for your needs. Follow the onscreen instructions to get everything set up. It won't take long.

>>>*Go Further*

TURNING ON AND CONFIGURING A NEW PC—FOR THE FIRST TIME

The first time you power up your new PC is different from what happens after you have everything set up. It's a more involved process because Windows walks you through a configuration process that gets your computer ready for you to use.

When you first turn on your new PC (by pressing the computer's "on" or power button), Windows displays a series of setup screens. You're asked a number of questions that are used to properly configure Windows for your use. For example, you need to select the region where you live, the language you speak, and so on. You also select your Wi-Fi network and enter the appropriate password.

During this initial setup process, you need to enter the email address and password for your Microsoft account. If you don't have a Microsoft account, click Create Account and follow the onscreen instructions.

Throughout this entire process, just follow the onscreen instructions and make the necessary choices. When you're done, Windows finishes the installation process and displays the desktop, with everything set up and ready to use.

The *next* time (and all subsequent times) you turn on your computer, things are a lot simpler, as noted in the following steps.

Booting Up

Technical types call the procedure of starting up a computer *booting* or *booting up* the system. Restarting a system (turning it off and then back on) is called *rebooting*.

Turn On Your Computer

After you've gone through the initial setup and configuration, turning on your computer is easy, especially if you have a notebook or 2-in-1 PC. It's just a matter of powering on everything connected to your computer—in the right order.

(1) Turn on your printer, monitor (for a traditional desktop PC), and other powered external peripherals.

(2) If you're using a laptop PC, open the laptop's case so that you can see the screen and access the keyboard.

(3) Press the power or "on" button on your computer. Windows launches automatically and displays the lock screen.

Lock Screen Information

The Windows lock screen displays a photographic background along with some useful information—including the date and time, power status, and Wi-Fi (connectivity) status.

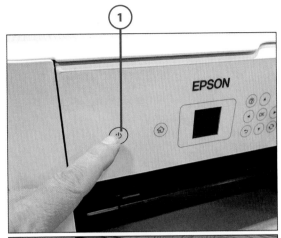

(4) Press any key or move your mouse to display the sign-in screen.

(5) Enter your password or PIN or use your PC's fingerprint scanner—however you've configured your computer's security—and then press the Enter key on your keyboard or click the next arrow key onscreen. Windows displays the desktop, ready for use.

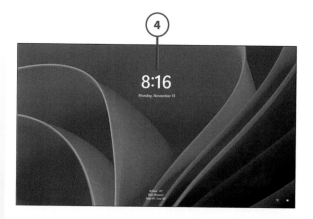

Log In Options

Learn more about your PC's log-in options in Chapter 7, "Working with Different Users."

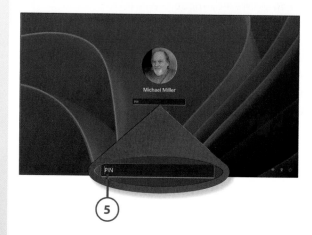

Turn Off Your Computer

How you turn off your PC depends on what type of computer you have. If you have a laptop or 2-in-1, you can press the unit's power (on/off) button—although that typically puts your PC into Sleep mode instead of turning it all the way off. The better approach is to shut down your system through Windows.

(1) Click or tap the Start button on the taskbar or press the Windows key on your computer keyboard to display the Start menu.

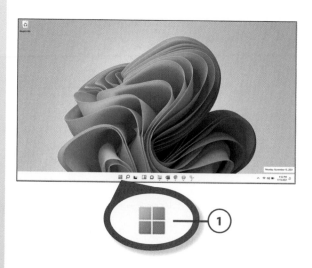

2 Click or tap the Power icon to display the submenu of options.

3 Click or tap Shut Down.

Sleep Mode

If you're using a laptop or 2-in-1 PC, Windows includes a special Sleep mode that keeps your computer running in a low-power state, ready to start up quickly when you open the lid or turn it on again. You can enter Sleep mode from the Power Options menu—or, with many laptop PCs, by pressing the unit's power button. (There's also a *hybrid sleep* mode available only on desktop PCs that places any open documents in memory while the PC goes into a low-power state.)

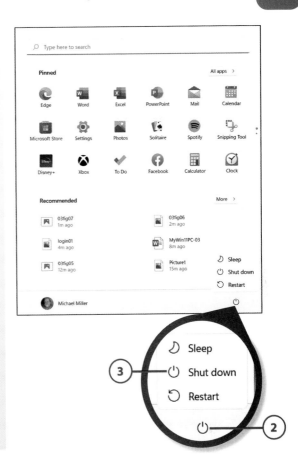

Finding Your Way Around Windows

When it comes to finding your way around Windows 11, it's all about learning the different parts of the desktop. (And, for you more experienced users, the Windows 11 desktop looks and works a bit differently than what you're used to with Windows 10.)

Use the Start Menu

You access all the software programs
and utilities on your computer via the
Windows Start menu. You can "pin"
your favorite programs to the Start
menu and view apps and files recommended by Windows. You can even
view all the apps installed on your
computer and power off your computer from the Start menu.

1. Click or tap the Start button or
   press the Windows key on your
   keyboard to open the Start menu.

2. At the top of the Start menu, you
   see apps that have been pinned
   to the Start menu. Mouse over
   the Pinned area and scroll down
   or back up to view all the pinned
   apps. Click or tap any app to
   open it.

3. At the bottom of the Start menu,
   you see recommended apps and
   recently used files. Click or tap an
   app or file to open it.

4. View all the apps installed on
   your computer by clicking or tapping All Apps.

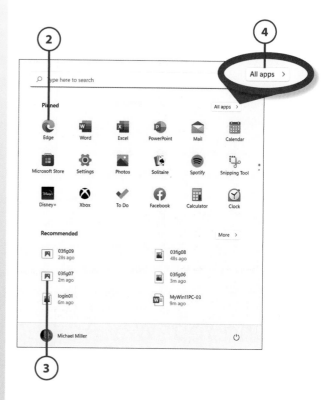

Pinning Apps

Learn more about pinning apps to the
Start menu and taskbar in Chapter 6,
"Personalizing Windows."

---

(5) Apps are listed in alphabetical order. Scroll down the list to view more apps.

(6) Click or tap an app to open it.

(7) Click or tap Back to return to the main Start menu.

Search for Apps and Files

You can also search for specific apps and files from the Start menu. Just click or tap within the Search field at the top of the Start menu to display the Search pane. (Learn more about searching from the Search pane in Chapter 8, "Using Apps and Programs.")

(8) View more recently used files by tapping More in the Recommended section.

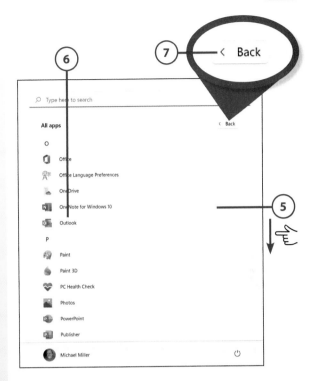

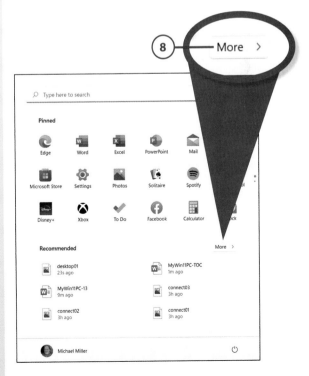

9 Recently opened files are listed in reverse chronological order, most recent files first. Scroll down to view older files.

10 Click or tap any file to open that file in the appropriate app.

11 Click or tap Back to return to the main Start menu.

12 Click or tap your account name or picture to change account settings, lock your PC, or sign out of your account.

13 Click or tap the Power icon to put your PC to sleep, shut it down, or restart it.

14 Close the Start menu by clicking or tapping anywhere outside the Start menu or clicking or tapping the Start button again.

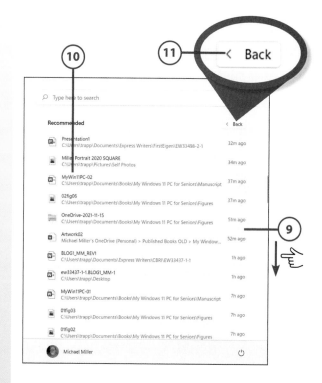

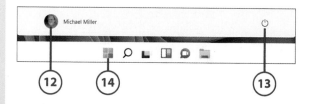

Different Looks

Your Start menu probably looks a little different from the ones shown in this chapter—in particular, the icons you see. That's because every person's system is different, depending on the particular programs and apps you have installed on your PC.

Quick Access Menu

If you right-click (instead of left-click) the Start button, you'll display an alternate Quick Access menu. This is a menu of advanced options, including direct links to File Explorer, Mobility Center, and Task Manager.

Use the Taskbar

The taskbar is that area at the bottom of the Windows desktop. Icons on the task-bar can represent frequently used programs, open programs, or open documents.

Centered Taskbar Icons

In Windows 11, the main taskbar icons are centered by default. (If you've used previous versions of Windows, you're probably used to them being on the left side of the taskbar.) If you'd rather move the icons to the left side of the taskbar, read Chapter 6 for more information.

(1) Open the Start menu by clicking or tapping the Start button.

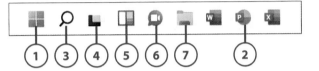

(2) Open any application pinned to the taskbar by clicking or tapping the application's icon.

(3) Search your computer for files and apps, or the Web for additional information, by clicking or tapping the Search icon. This opens the Search pane, described in Chapter 8.

(4) View all open applications in thumbnail form by clicking or tapping the Task View button. (Learn more about switching between programs in Chapter 8.)

(5) Open the Widgets pane by clicking or tapping the Widgets button. (Learn more about widgets in Chapter 6.)

(6) Start or join a Microsoft Teams chat by clicking or tapping the Chat icon.

(7) Open File Explorer by clicking or tapping the File Explorer icon.

8 The far-right side of the taskbar is called the notification area, and it displays icons for essential Windows operations—sound, networking, power, time and date, and so forth. Click or tap the up arrow to view icons for more items, normally hidden.

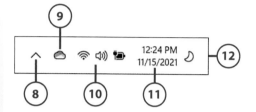

9 Open OneDrive by clicking or tapping the OneDrive icon.

10 Connect to Wi-Fi networks, adjust sound and brightness levels, and make other quick adjustments by clicking or tapping the middle of the notification area to display the Quick Access panel.

11 View recent notifications from Windows and selected apps by clicking or tapping the time and date.

12 Minimize all open applications by clicking or tapping the slim Peek button at the far right of the taskbar.

Taskbar Icons

A taskbar icon with nothing underneath represents an unopened application. A taskbar icon with a short line underneath represents a running application. A taskbar icon with a longer line underneath and a slightly shaded background represents the highlighted or topmost window on your desktop.

Use the Quick Settings Panel

Windows 11 features a new Quick Settings panel that you use to adjust basic settings—changing volume and brightness levels, connecting to Wi-Fi networks, switching to Airplane mode, and adjusting the screen brightness. You open the Quick Settings panel from the notification area of the taskbar.

(1) Click or tap the middle of the notifications area on the taskbar to open the Quick Settings panel.

(2) Use the Brightness slider to adjust the screen brightness.

(3) Use the Volume slider to adjust the audio volume.

(4) Click or tap the Wi-Fi button to turn Wi-Fi on or off.

(5) Click or tap the Wi-Fi right arrow to connect to or switch Wi-Fi networks.

(6) Click or tap the Airplane Mode button to turn Airplane mode on or off.

(7) Click or tap the Settings button to open the Settings app.

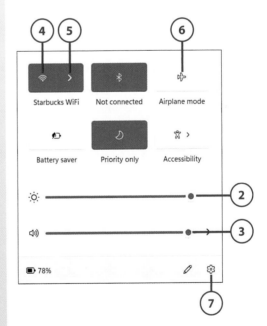

Use the Notifications Panel

The new Notifications panel in
Windows 11 is where you view system
notifications and notifications from
selected apps. It also displays a handy
calendar.

1. Click or tap the date and time
   area of the notifications area
   of the taskbar to display the
   Notifications panel.

2. Recent notifications are displayed
   here. Scroll down to view more.

3. Click or tap to read or take action
   on any specific notification.

4. Mouse over any notification
   and click the X to close that
   notification.

5. Click or tap Clear All to close all
   notifications.

6. Click or tap the up arrow by
   the date to display a monthly
   calendar. The calendar expands
   up while the notifications panel
   contracts upward.

7. Click or tap the up and down
   arrows to display the previous
   and upcoming months.

8. Click the down arrow next to
   the calendar to minimize the
   calendar and display more
   notifications.

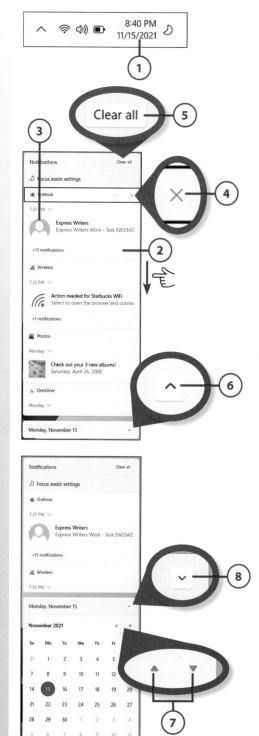

Switching from Windows S Mode to Windows Home

If you just purchased a new PC, it's likely that your computer is running a special version of Windows called Windows S Mode. Windows S Mode is just like regular Windows except it can run only Windows apps downloaded from the Microsoft Store. Computers running in Windows S Mode cannot use traditional desktop software, such as Adobe Reader and Photoshop Elements.

Cloud Apps in S Mode

Windows S Mode can run cloud-based apps that run within a web browser, such as Google Docs.

Microsoft's stated reason for introducing Windows S Mode is to make Windows-based computers more secure. Apps available in the Microsoft Store are "Microsoft-verified" for security, whereas traditional software apps are not.

The built-in limitations of Windows S Mode make it less than ideal for many computer users, especially those using older software not available in the Microsoft Store. Fortunately, you can quickly and easily switch your version of Windows from Windows S Mode to Windows Home, which does run traditional desktop software. The switch takes just a few minutes of your time, and it's totally free.

Is Your Computer in S Mode?

To see if your computer is running in S Mode, click the Start button and select Settings. From the Settings app, click System and then select the About tab. Scroll to the Windows Specifications section and look at the Edition entry. If it says Windows 11 Home in S Mode, your computer is running in S Mode. If it doesn't say S Mode, it's not.

Switch from Windows S Mode to Windows Home

Any computer running Windows S Mode, new or old, can be upgraded to Windows Home for free. (Note, however, that you cannot switch back to Windows S Mode from Windows Home; this is a one-way switch.)

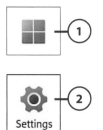

(1) Click or tap the Windows Start button to open the Start menu.

(2) Click or tap Settings to open the Settings tool.

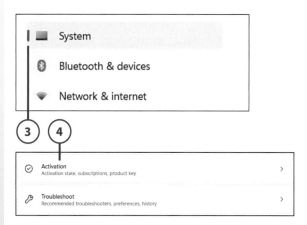

(3) Click or tap System on the left.

(4) Scroll down and click or tap Activation on the right.

(5) Go to the Switch to Windows 11 Home section and click or tap Go to the Store. This opens the Microsoft Store app to the appropriate Switch Out of S Mode page.

(6) Click or tap the Get button.

(7) It only takes a few seconds— no rebooting required—for your computer to switch from Windows S Mode to Windows Home. When Windows notifies you that the switch is complete, click or tap Close.

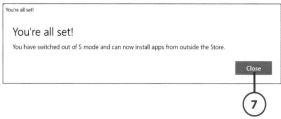

>>>*Go Further*

WINDOWS HOME, PRO, ENTERPRISE, AND EDUCATION—AND S MODE

Microsoft sells four primary versions of Windows 11: Windows Home, Windows Pro, Windows Enterprise, and Windows Education. At first glance, it's difficult to tell the differences between them, although the differences are there.

Most home and small business computers run Windows Home, whereas larger businesses and organizations run Windows Pro or Windows Enterprise. The Pro and Enterprise versions are functionally identical to Windows Home but offer more business-specific security and data management features.

There's also a Windows Education version, designed for use in schools. This version is similar to Windows Enterprise with its enhanced security and network management functions.

S Mode is available for all four of these versions. So, a home computer may be running Windows Home in S Mode and a business computer may be running Windows Pro in S Mode. When you switch out of S Mode, you switch to the main version of Windows (Home, Pro, or otherwise) installed on your computer.

In this chapter, you find out how to use
Windows 11 if you've previously used Windows
10 or another older version of Windows.

4

Windows 11 for Windows 10 Users

The Windows operating system has been around for more than 35 years now. Version 1.0 of Windows was released in November of 1985 and has gone through numerous small and more significant revisions since then.

Windows 11, released in October 2021, is the latest version of Windows—and the first major upgrade since Windows 10 was released more than six years ago.

What's new in Windows 11? The answer is a lot. Read on to learn more.

Can Your PC Run Windows 11?

All new PCs sold today come with the Windows 11 operating system preinstalled. If you have an older computer, however, you may want to upgrade from Windows 10 or another previous version of Windows to the newer Windows 11 operating system.

The question is, can you install Windows 11 on your existing computer? It depends on whether your computer meets the following specifications:

- 8th-generation or newer Intel CPU or AMD Ryzen 2000 or newer CPU
- 4 GB or more of RAM
- 64 GB or larger hard drive or SSD storage
- Graphics card compatible with DirectX 12 or later with WDDM 2.0 driver
- High-definition monitor or display with at least 720p (1280×720) resolution and 9" or larger screen
- Internet connection
- Microsoft account (if you don't have one, you can create one during initial setup)
- System firmware that is UEFI and Secure Boot capable
- TPM 2.0 cryptoprocessor (common on computers manufactured after 2016)

These are, by the way, much stricter requirements than Microsoft had to run Windows 10. If your computer doesn't meet these requirements, it won't be able to run Windows 11. Your options are to continue using your old computer with Windows 10 (which is still supported by Microsoft) or buy a new PC with Windows 11 preinstalled.

PC Health Check

If you're not sure whether your current computer meets the specifications to run Windows 11, download and install Microsoft's PC Health Check app. When you run PC Health Check, it tells you whether you can run Windows 11—and if not, why. Download it from Microsoft's main Windows 11 web page, located at www.microsoft.com/en-us/windows/windows-11.

If your computer meets the requirements, Microsoft offers the Windows 11 upgrade to you—eventually. It takes time to update the hundreds of millions of Windows computers, so Microsoft staggers the upgrade process. You'll get an onscreen notification when your computer is ready to be upgraded; just follow the instructions to download and install Windows 11.

What's New in Windows 11?

Chances are you're not new to computing in general or Windows in particular. If you've used Windows 10 or any previous version of Windows, Windows 11 looks quite familiar—while at the same time, subtly different. That's how Microsoft planned it.

What new and changed things will you find in Windows 11? Here's a short list:

- Updated interface with rounded corners and pastel colors
- Centered Start menu on the taskbar
- Revamped Start menu with icons instead of tiles
- Revised interface in the Settings app
- Revised interface in File Explorer
- Action Center broken into separate Quick Settings and Notifications panels
- Ability to run select Android apps from within Windows (not available at launch; will be added on a later update)
- Personalizable widgets accessible directly from the taskbar
- Microsoft Teams integration, available directly from the taskbar (replaces Skype integration in Windows 10)
- Updated virtual desktop support and functionality
- Snap Groups and Snap Layouts for easier multitasking
- New docking and undocking functionality for multiple monitor use
- Redesigned interface for tablet and 2-in-1 use, with more space between icons
- Redesigned Microsoft Store, now including desktop apps
- Updated Xbox technology for better PC gaming, including Auto HDR and DirectStorage

You see a lot of these changes when you first boot up Windows 11. The desktop experience is noticeably different with the centered Start menu, rounded window corners, and redesigned icons. It's more streamlined and modern looking, which is a good thing.

The new Windows 11 desktop and interface

>>>Go Further
WINDOWS UPDATES

Although Windows 11 is the first new version of Windows since 2015, it's not the first update to Windows 10. That's because Microsoft released an average of two major Windows 10 updates every year, along with smaller monthly updates. In fact, there were nine major updates to Windows 10 over the course of its lifetime.

Will Windows 11 have the same types of monthly and biannual updates as Windows 10 did? Yes, but slightly different. Expect to receive a relatively small update (mainly bug fixes) once a month and a larger update (often with new features and functionality) just once a year. All these updates should be automatic and occur in the background, so they won't interrupt your work.

How to Do the Same Old Things—the New Windows 11 Way

The good news about Windows 11 is that even though it may look a little different (and have a few new features), most basic operations are pretty much the same as they were in Windows 10. There are some things, however, that you do differently, as detailed in Table 4.1.

Table 4.1   Changed Operations in Windows 11

Operation	Windows 10 Way	Windows 11 Way
Access key system settings and notifications	Open Action Center.	Action Center replaced by separate Quick Settings and Notifications panels.
Log in to Wi-Fi networks	Click the Connections icon in the Notifications area to display the Connections panel and then select a wireless network; you might need to enter a password or open your web browser to log into a public network.	Click the left side of the Notifications area to display the Quick Settings panel; then click the right arrow next to the Manage Wi-Fi Connections button and select a wireless network. If the network requires logging in, click the More Actions link.
Open Settings app	Click the Notifications button and select All Settings.	Click the Start button and select Settings.
Open Task Manager	Right-click the taskbar and select Task Manager.	Right-click the Start button and select Task Manager.
Switch to Tablet mode	Switched automatically.	No separate Tablet mode; interface slightly changes when PC is used as a tablet.
View short bits of information	Live tiles on Start menu.	Widgets on Widget panel.
Video chat	Use Skype.	Use Microsoft Teams.

USB Type-A connector

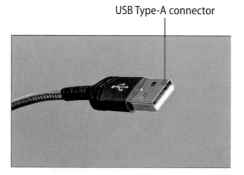

USB Type-C connector

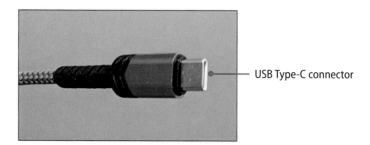

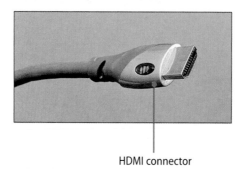

HDMI connector

In this chapter, you find out how to connect your new computer to printers and other USB devices.

Connecting Printers and Other Peripherals

Your Windows 11 computer can't do everything itself. To get the most out of your machine, you might want to connect it to other devices—such as a printer or even your living room TV.

Connecting Devices via USB

Most external devices connect to your PC via USB. This is a type of connection common on computers and other electronic devices; it carries data and provides power for some connected devices.

USB is popular because it's so easy to use. All you have to do is connect a device via USB and your computer should automatically recognize it.

USB

USB stands for *universal serial bus* and is an industry standard developed in the mid-1990s. On today's computers, you may find a mix of traditional USB Type-A ports and the smaller USB Type-C ports that are common to smartphones and other mobile devices.

Connect a Peripheral Device

You can connect a variety of peripherals to your computer via USB. These include mice, keyboards, digital cameras, external hard drives, and scanners.

(1) Connect one end of a USB cable to your new device.

(2) Connect the other end of the cable to a free USB port on your PC.

(3) In most cases, Windows recognizes new devices and automatically installs the proper system drivers and files. If Windows can perform multiple actions for a given device (such as viewing or downloading photos from a digital camera), you might be prompted to select which action you want to take. Click or tap the prompt to make a selection.

USB Hubs

If you connect too many USB devices, you can run out of USB connectors on your PC. If that happens, you can buy an inexpensive add-on USB hub, which lets you plug multiple USB peripherals into a single USB port.

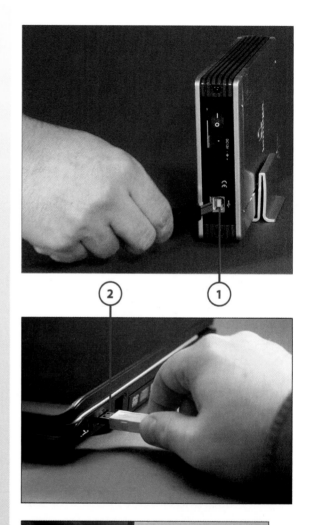

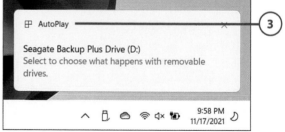

Connecting a Printer

When you want to make a physical copy of something onscreen—a letter, a spreadsheet, a photo—you need to connect a printer to your computer. You can connect both inkjet and laser printers, as well as multifunction printers that offer scanning, copying, and even faxing features. Most printers can connect either wirelessly (via Wi-Fi) or via a USB connection.

Connect a Wireless Printer

Most printers today are wireless printers, meaning that they connect to your computer wirelessly via Wi-Fi. This lets you place your printer anywhere in your house without having to physically tether it to your PC. (It also makes it easy to share a single printer with multiple computers and other devices.)

1. Power on your printer and follow the manufacturer's instructions to connect it to your Wi-Fi network.

2. On your computer, click or tap the Start button to open the Start menu.

3. Click or tap Settings to display the Settings window.

4. Click or tap Bluetooth & Devices.

5. Click or tap Printers & Scanners.

6. Click or tap Add Device and let Windows search for your wireless printer.

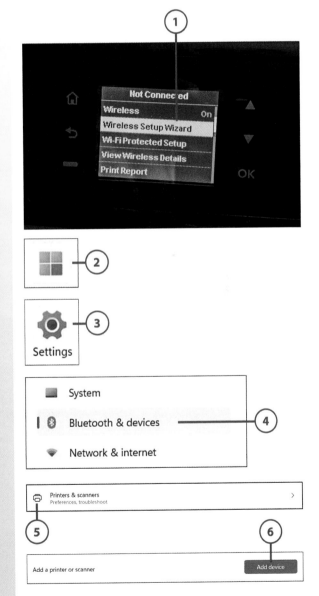

(7) Your wireless printer should appear in the resulting list. Select it and then click or tap Add Device. Follow the onscreen instructions to complete the installation.

| | HP ENVY 4500 series [EBF3D1] Scanner | Add device |

It's Not All Good

Not Always Reliable

When working with printers, I've sometimes found USB connections to be more reliable than wireless ones, although your experience may vary. Sometimes a computer can lose the wireless connection to the printer, which forces you to reboot both the printer and computer. USB connections seldom suffer this problem.

Connect a Printer via USB

If your printer is physically close to your computer, it may be easier to connect your printer via a USB cable. The setup is certainly easier.

(1) Connect one end of a USB cable to the USB port on your printer. (Some printers have a USB Type-B port, which requires a USB Type-B–to–USB Type-A cable to connect to your computer.)

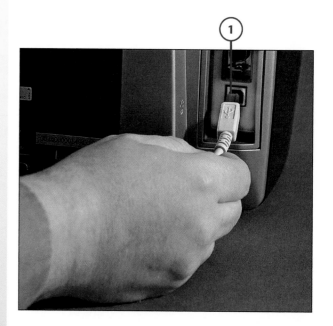

(2) Connect the other end of the USB cable to a USB port on your computer.

(3) In many cases, Windows recognizes the printer and installs it automatically. If this doesn't happen, you can install the printer manually. Begin by clicking or tapping the Start button to open the Start menu.

(4) Click or tap Settings to display the Settings window.

(5) Click or tap Bluetooth & Devices.

(6) Click or tap Printers & Scanners.

(7) Click or tap Add Device and let Windows search for your printer.

(8) Your printer's name should appear in the resulting list. Select it and click Add Device. Then follow the onscreen instructions to complete the installation.

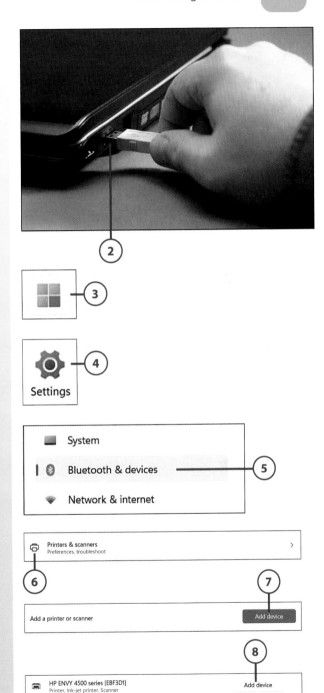

System

Bluetooth & devices —— (5)

Network & internet

🖶 Printers & scanners
Preferences, troubleshoot   >

(6)   (7)

Add a printer or scanner   Add device

(8)

HP ENVY 4500 series [EBF3D1]   Add device
Printer, Ink-jet printer, Scanner

Connecting Your PC to Your TV

If you want to watch Internet streaming video (from Amazon Prime, Netflix, YouTube, and other services) on your TV, you can simply connect your TV to your PC via an HDMI cable. After you connect your TV and PC this way, anything you watch on your PC displays on your TV screen.

Connect via HDMI

HDMI is the easiest way to connect your PC to your TV. HDMI stands for *high-definition multimedia interface,* and it has become the connection standard for high-definition TVs. All newer TV sets have two or more HDMI inputs, which you typically use to connect cable boxes, Blu-ray players, and the like. HDMI transmits both audio and video signals.

Most new computers—both desktops and laptops—have either a full-sized or mini HDMI port. All you have to do is connect the appropriate HDMI cable between your two devices.

1. Connect one end of an HDMI or mini HDMI cable to the HDMI port on your computer.

2. Connect the other end of the HDMI cable to an open HDMI connector on your TV.

(**3**) Switch your TV to the HDMI input you connected to.

(**4**) On your computer, press Win+P on the keyboard to display options for the external display. Click or tap Duplicate to display content on both your computer screen and the TV screen. *Or…*

(**5**) Click or tap Second Screen Only to display content only on the TV screen while the computer screen is blank.

Mini HDMI Connectors

Not all PCs have full-sized HDMI ports. Some laptops have mini HDMI connectors, which require the use of a special HDMI cable with a mini connector on one end and a standard-sized connector on the other.

(**3**)

(**4**)

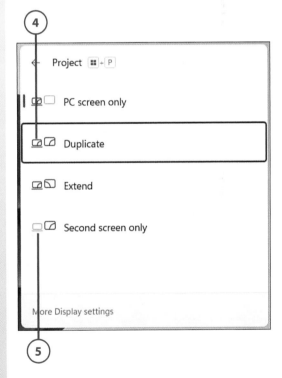

(**5**)

Wirelessly Mirror Your Computer Screen

If you have a so-called smart TV or have a streaming media player (such as Amazon Fire TV or Roku) connected to your TV, you can wirelessly mirror the contents of your computer screen to that TV through a technology called Miracast. You just have to have your TV connected to the same Wi-Fi network as your computer.

1. Enable the screen mirroring feature on your TV or streaming media player.

2. On your computer, press Win+K to display screencasting options. Click or tap to select the name of the device you want to cast to. Follow any onscreen instructions to complete the connection.

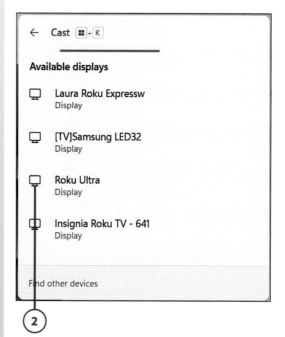

Windows 11 desktop and Start
menu in light mode

Windows 11 desktop and Start
menu in dark mode

Personalizing Windows

When you turn on your computer, you see the Windows lock screen and then the Windows desktop. You can accept the default look for each of these items, or you can customize them to your taste. It's one way to make Windows look like *your* version of Windows—and make your workplace more efficient.

Personalizing the Start Menu and Taskbar

You can personalize which app icons appear on the taskbar and on the Pinned section of the Start menu. This helps you get to your favorite apps faster.

Personalize the Start Menu

The Windows 11 Start menu is different from the Start menu in Window 10 in that it displays small icons for your favorite and most-used apps instead of the larger tiles in the older operating system. This makes for a leaner and more efficient Start menu experience that you can still customize by "pinning" icons for your favorite apps.

Pinning

"Pinning" an app creates a shortcut to that app. You can pin programs to either the Start menu or the taskbar. Pins you add can be removed at any time.

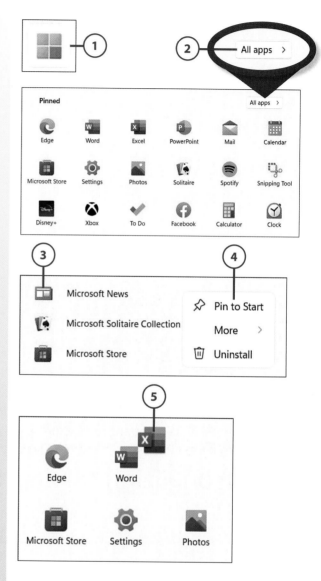

1. Click or tap the Start button to open the Start menu.

2. "Pin" a program to the top half of the Start menu by clicking or tapping All Apps to display all the apps installed on your computer.

3. Right-click the icon for the app you want to pin.

4. Click or tap Pin to Start.

5. Rearrange the icons for your pinned apps by clicking and holding an icon and then dragging it to a new position.

6. Move an icon to another page of pinned apps by clicking and dragging the icon to the top or bottom of the Pinned area until you see the previous or next page.

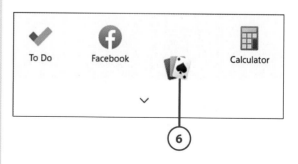

(7) Remove a pinned app from the Start menu by right-clicking the icon and selecting Unpin from Start.

Personalize the Taskbar

You can personalize the way the taskbar appears. You can opt to shift the icons to the left side of the taskbar (instead of centering them) and select which icons appear in the notification area.

(1) Right-click any open area of the taskbar and then click or tap Taskbar Settings to see the four system icons you can choose to show or not show on the taskbar: Search, Task View, Widgets, Chat.

(2) Click or tap each item's switch "on" to show it on the taskbar or click or tap it "off" to hide it.

(3) You can also opt to show three additional icons in the corner of the taskbar: Pen Menu, Touch Keyboard, and Virtual Touchpad. Click or tap each item's switch "on" to show it on the taskbar or click or tap it "off" to hide it.

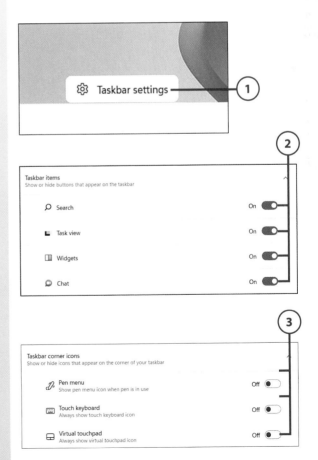

(**4**) There are several additional icons you can display in the corner of the taskbar by the notification area, including Microsoft OneDrive, Windows Explorer, Windows Security Notification Icon, Microsoft Outlook, and Windows Update Status. Click or tap to expand the Taskbar Corner Overflow section; then click or tap each item "on" or "off."

(**5**) Change the alignment of the taskbar icons by clicking or tapping to expand the Taskbar Behaviors section. Then pull down the Taskbar Alignment list and select either Center (default) or Left.

(**6**) Hide the taskbar when you're not using it (it appears when you mouse over the bottom of the screen) by clicking or tapping to select Automatically Hide the Taskbar.

(**7**) Display a number counter for unread messages (called "badges") on select app icons by clicking or tapping to select Show Badges. This option is on by default; deselect this option to hide those little numbers.

(**8**) To be able to click or tap the far-right corner of the desktop to minimize all open windows and show the desktop, click or tap Select the Far Corner of the Desktop to Show the Desktop.

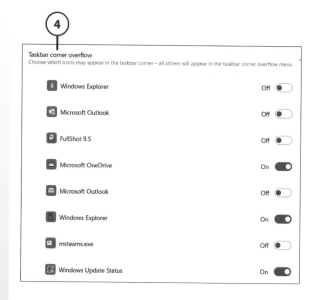

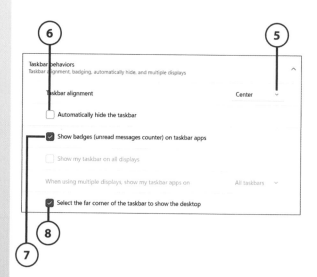

You Can't Move It

In previous versions of Windows, you could drag the taskbar to the top or either side of the screen. You can't do that in Windows 11; the new taskbar is immovable. (This also means you don't have to lock it in position, so there's no option for that anymore, either.)

Pin Apps to the Taskbar

Just as you can pin your favorite apps to the Start menu, you can pin apps to the taskbar. Click on an icon for a pinned app to open that app.

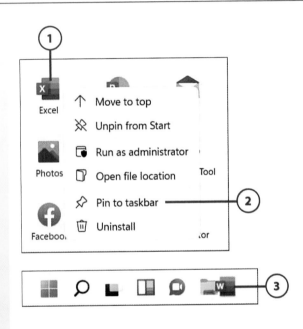

1. From the Start menu, navigate to the app you want to pin to the taskbar.

2. Right-click the app icon and select Pin to Taskbar.

3. Rearrange pinned apps on the taskbar by clicking and dragging an icon left or right to a new position.

>>>Go Further

CREATE APP SHORTCUTS ON THE DESKTOP

Although you can't pin apps to the desktop, you can create nearly identical shortcuts to those apps on the desktop. The process for creating a desktop shortcut is a little different from pinning an app to the Start menu or taskbar, however.

Start with a clean desktop, with all windows minimized. Next, open the Start menu and click All Apps. (This step is important; you can't create a shortcut from the Start menu's Pinned section.)

Scroll to the app you want and then click and drag it off the Start menu onto the desktop. This creates a shortcut to that app on the desktop, and you see the app icon in two places: the Start menu and the desktop.

Personalizing the Windows Desktop

You can personalize several elements on the Windows 11 desktop. You can change the color scheme, choose your desktop background, and switch from Light to Dark mode.

Change the Desktop Background

The Windows desktop displays across your entire computer screen. One of the most popular ways to personalize the desktop is to use a favorite picture or color as the desktop background.

1. Right-click in any open area of the desktop and select Personalize from the pop-up menu. The Personalization tab of the Settings app displays.

2. Click or tap Background.

3. Use a picture as your desktop background by clicking the Personalize Your Background control and selecting Picture.

4. Click to select one of the recent images displayed. *Or...*

5. Click Browse Photos to select another picture stored on your computer.

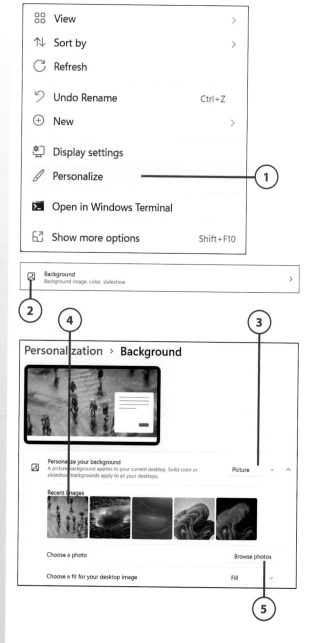

6 If the selected image is a different size than your Windows desktop, click the Choose a Fit for Your Desktop Image list and select a display option—Fill (zooms into the picture to fill the screen), Fit (fits the image to fill the screen horizontally, but might leave black bars around the image), Stretch (distorts the picture to fill the screen), Tile (displays multiple instances of a smaller image), Center (displays a smaller image in the center of the screen, with black space around it), or Span (spans a single image across multiple monitors, if you have multiple monitors on your system).

7 Set a color for your desktop background by clicking the Personalize Your Background list and selecting Solid Color.

8 Click to select the color you want or click View Colors in the Custom Colors section to choose from a broader palette.

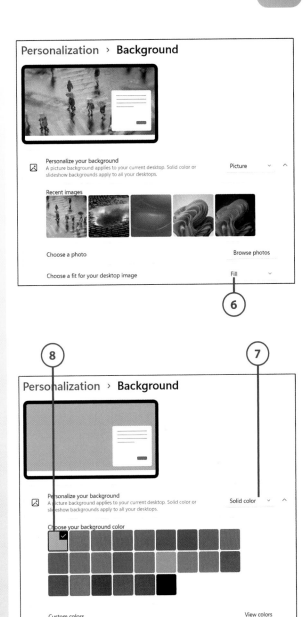

9 To have your desktop background rotate through a variety of pictures, click the Personalize Your Background list and select Slideshow.

10 By default, the slideshow chooses pictures from your Pictures folder. To select a different folder, click Browse.

11 Change how long each photo is displayed by clicking the Change Picture Every list and making a new selection.

12 Display pictures randomly by clicking "on" the Shuffle switch.

13 By default, the desktop slideshow pauses if your laptop computer is running on battery power. To continue the slideshow when your computer isn't plugged in (which drains your battery faster), click "on" the Let Slideshow Run Even If I'm on Battery Power switch.

14 Click the Choose a Fit list and select a display option.

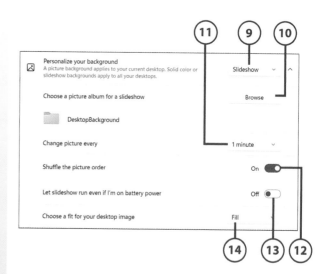

Change the Accent Color

You can select any color for the title bar and frame that surrounds open windows on the desktop. You can also set the color for the Windows taskbar, Start menu, and Action Center.

1 Right-click in any open area of the desktop and select Personalize from the pop-up menu. The Personalization tab of the Settings app displays.

2 Click or tap Colors.

3 Make the Windows desktop elements transparent by clicking "on" the Transparency Effects switch.

4 To have Windows automatically choose the accent color based on the color of the desktop image, click the Accent Color control and select Automatic.

5 To manually select an accent color, click the Accent Color control and select Manual; then click to select the color you want.

6 If you're in Windows Dark mode (discussed next), you can show the accent color on the Start menu and taskbar by scrolling down the window and clicking "on" the Show Accent Color on Start and taskbar switch. Click this switch "off" to show these elements in standard system colors (gray in Light mode, black in Dark mode).

7 Show the accent color on windows title bars and borders by clicking "on" the Show Accent Color on Title Bars and Window Borders switch. Click this switch "off" to display these elements in standard system colors.

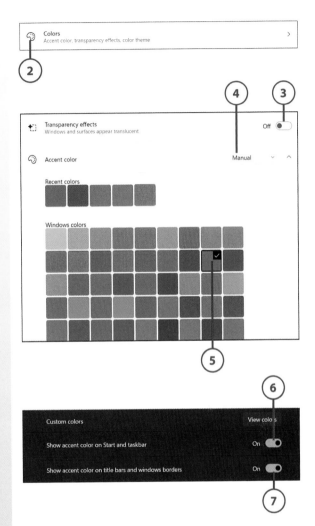

Switch to Dark or Light Mode

Windows 11 offers both a Dark mode and Light mode. Dark mode displays a black background in the taskbar, Start menu, and many windows. Light mode displays a light gray background in these same areas.

1. Right-click in any open area of the desktop and select Personalize from the pop-up menu. The Personalization tab of the Settings app displays.

2. Click Colors.

3. Click the Choose Your Mode list and select either Light or Dark.

4. Set one mode for Windows elements and the other for apps by selecting Custom. This expands the Choose Your Mode section.

5. Click the Choose Your Default Windows Mode list and select either Light or Dark to set the mode for the Start menu, taskbar, and other Windows elements.

6. Click the Choose Your Default App Mode list and select either Light or Dark to set the mode for your Windows apps.

Custom Mode

You may like the Dark mode for Windows elements but not like a dark background in Microsoft Word and other apps. In this instance, select Custom and then select Dark mode for the first option and Light mode for the second.

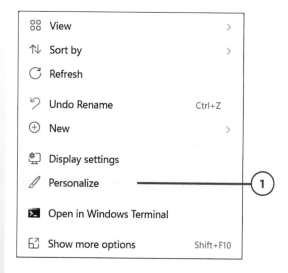

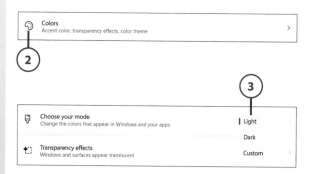

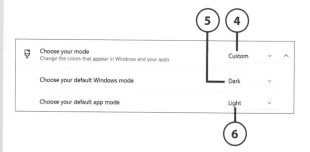

Change the Desktop Theme

Although you can configure each element of the Windows desktop separately, it's often easier to choose a predesigned *theme* that changes all the elements in a visually pleasing configuration. A theme combines background images, color schemes, system sounds, and mouse cursor appearance to present a unified look and feel. Some themes even change the color scheme to match the current background picture.

1. Right-click any open area of the desktop to display the options menu and then click Personalize to display the Personalization tab of the Settings app.

2. Click Themes.

3. To save the currently selected background, color, sound, and mouse scheme as a new theme, click Save. When prompted, give this new theme a name.

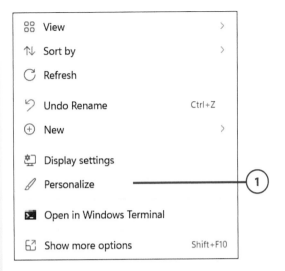

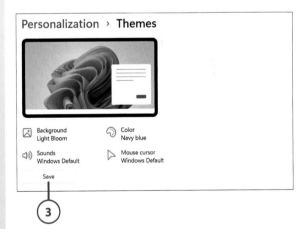

4 Scroll down to the Current Theme section to view all themes installed on your PC. Click any theme to change to that theme.

5 Additional themes, most free of charge, are available from the Microsoft Store online. Click Browse Themes to view what's available and download new themes.

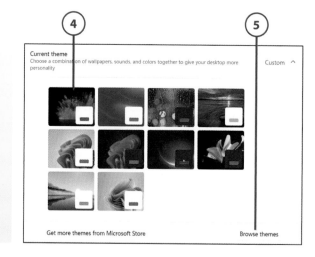

Personalizing Widgets

A widget is a small panel that displays specific information, such as news, weather, appointments, and the like. Many apps, such as Photos and Weather, have their own related widgets. Other widget content is sourced from around the Web.

Display and Use the Widgets Panel

Widgets are displayed in a Widgets panel that slides in from the left side of the desktop. There are two ways to open the Widgets panel:

1 Click or tap the Widgets icon in the taskbar. *Or…*

2 On a touchscreen display, swipe in from the left side of the screen.

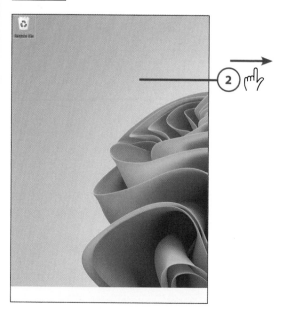

③ Scroll down to view more
   widgets.

④ Click the title of a widget to open
   the associated app or view more
   information.

⑤ To close the Widgets panel, click
   or tap anywhere else on the
   screen.

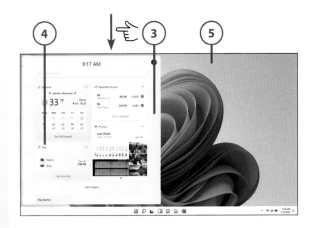

Personalize the Widgets Panel

You can personalize the Widgets panel
by resizing, deleting, and rearranging
widgets.

① Resize a widget by clicking or
   tapping the More Options (three-
   dot) button and selecting a different size.

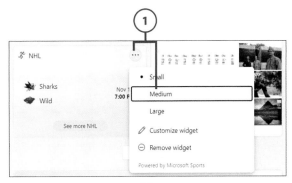

Widget Sizes

Not all widgets can be displayed at all
available sizes. Each widget displays in just
one of the two columns in the Widgets
panel; small widgets are shorter, whereas
medium and large ones are taller.

② Customize the content of a
   widget by clicking or tapping the
   More Options button and selecting Customize Widget.

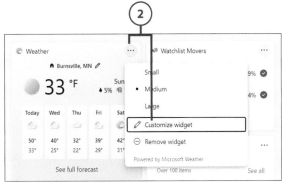

3 Each widget has its own customization options. For example, the Weather widget lets you set your location and whether to display temperatures in Fahrenheit or Celsius; the Sports widget lets you select which teams and sports to follow.

4 Rearrange widgets by using your mouse (or finger on a touchscreen device) to drag a widget to a different location.

5 Remove a widget by clicking or tapping the More Options button and selecting Remove Widget.

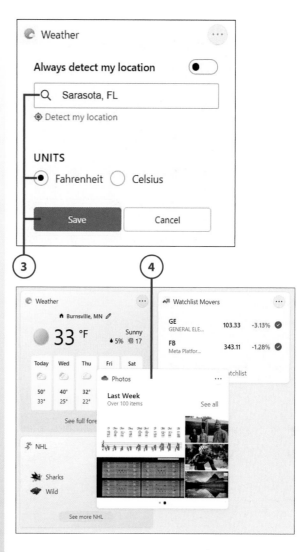

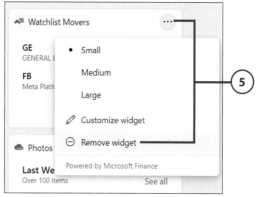

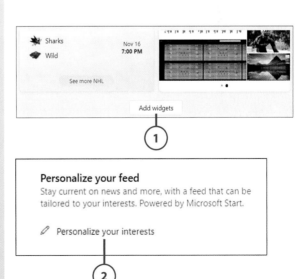

Add a New Widget to the Widgets Panel

Microsoft helpfully prepopulates the Widgets panel with a selection of popular widgets. You don't have to limit yourself to these particular widgets, however; it's easy to add new widgets to the panel at any time.

1 Scroll down the Widgets panel and click or tap Add Widgets.

2 Select the new widget(s) you want to add.

Personalize Your News Feed

As you scroll down the Widgets panel, you see tiles for news stories. These tiles are part of your News Feed, which is a permanent section of the Widgets panel. Click or tap the tile for any story to read the story in full in the Microsoft Edge web browser.

1 Scroll down the Widgets panel and click or tap Add Widgets.

2 In the Personalize Your Feed section, click or tap Personalize Your Interests. This opens Microsoft Edge to the My Interests page.

(3) Scroll down the list of interests and select those you're interested in.

(4) Deselect those interests in which you're not interested.

(5) Add a specific interest to your feed by entering that topic into the Discover Interests search box and then selecting the matching interest.

Less Like This

When you find a story that you don't like, mouse over that story's widget and click the X. You can then select that you're not interested in stories like this or you don't want to see any more stories from this particular source.

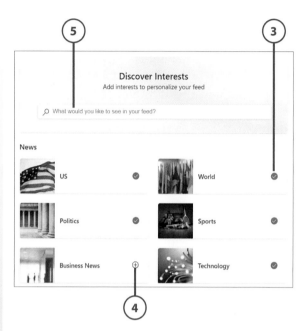

>>>Go Further

WIDGETS—THE NEXT GENERATION OF LIVE TILES

Widgets are only kind of new to Windows 11. Windows 10 had something similar—*live tiles* pinned to the Start menu.

That version of Windows had big tiles for the apps on the Start menu instead of app icons, and for certain apps, the tiles were "live"—that is, they displayed live content. For example, the tile for the Weather displayed the current temperature and weather conditions; the tile for the News app displayed a scrolling list of current headlines.

Because Windows 11 did away with the large and somewhat intrusive Start menu in favor of a more streamlined one, those live tiles no longer had a place to live, even though some users found them useful. The solution is widgets, which offer the same functionality as live tiles (and a little bit more) but in a different form factor.

Personalizing the Lock Screen

Another thing you can personalize is the lock screen, which you see when you first start or begin to log in to Windows. You can change the background picture of the lock screen, turn the lock screen into a photo slideshow, and add informational apps to the screen.

Change the Lock Screen Background

You can choose from several stock images for the background of your lock screen, or you can upload a photo to use as the background.

Lock Screen

The lock screen appears when you first power on your PC and any time you log off from your personal account or switch users. It also appears when you awaken your computer from Sleep mode.

1. Right-click in any open area of the desktop and select Personalize from the pop-up menu. The Personalization tab of the Settings app displays.

2. Click or tap Lock Screen.

3. Let Microsoft display different backgrounds for your lock screen by clicking the Personalize Your Lock Screen list and selecting Windows Spotlight.

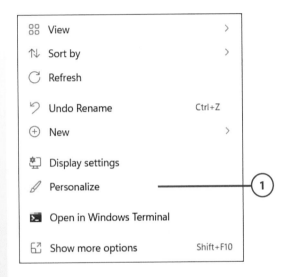

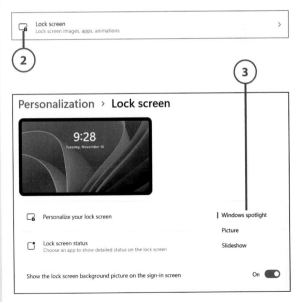

(4) Choose a specific picture for the lock screen by clicking the Personalize Your Lock Screen list and selecting Picture.

(5) Click the thumbnail for the picture you want to use. *Or...*

(6) Click the Browse Photos button to use a picture stored on your computer as the background.

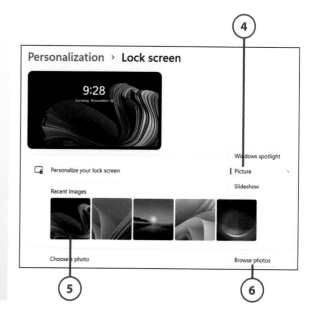

Display a Slideshow on the Lock Screen

Windows lets you turn your computer into a kind of digital picture frame by displaying a slideshow of your photos on the lock screen while you're not using your PC.

>>>*Go Further*

DISPLAY THE LOCK SCREEN

You can display the lock screen (and your photo slideshow) at any time by opening the Start menu, clicking your profile picture, and selecting Lock. You can also display the lock screen by pressing Win+L on your computer keyboard.

① From the Lock Screen section of the Settings app, click or tap the Personalize Your Lock Screen list and select Slideshow.

② By default, Windows displays pictures from your Pictures folder. Click the Browse button to select a different picture folder you want to use for your slideshow.

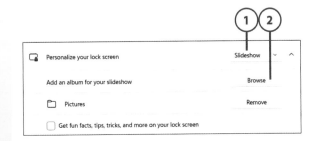

>>>Go Further
OTHER LOCK SCREEN SETINGS

To display what Microsoft calls "fun facts, tips, tricks, and more" on your lock screen, click or tap to select that option in the Lock Screen window. (It's selected by default, so uncheck it if you don't want to see those things.)

If you opted to display a slideshow on your lock screen, expand the Advanced Slideshow Settings section to see more options. Here you can configure the slideshow to

- Include Camera Roll folders from this PC and your OneDrive account
- Use only pictures that best fit your screen
- Not play a slideshow when on battery power
- Show the lock screen instead of turning off the screen when your PC is inactive

You can also choose to turn off your computer's screen after the slideshow has played for a specified period.

In addition, Windows lets you display one piece of live information on the lock screen. You can opt to see information from your Calendar, Mail, SupportAssist, Weather, or Xbox Console Companion app. (I like seeing weather information on the lock screen.) Just click the Lock Screen Status list and make a selection.

Finally, Windows displays the lock screen background picture on the sign-in screen by default. If you'd rather just see a blank background on the sign-in screen, click "off" the Show the Lock Screen Background Picture on the Sign-In Screen control.

Change Your Account Picture

Windows displays a small thumbnail image next to your account name when you log in to Windows from the lock screen; this same image displays next to your name on the Windows Start menu. When you first configured Windows, you were prompted to select a default image to use as this profile picture. You can, at any time, change this picture to something more to your liking.

1. Click the Start button to display the Start menu.

2. Click your name or picture at the bottom of the Start menu to display the options menu.

3. Click Change Account Settings to display the Settings app with the Your Info page selected.

4. Select a picture stored on your computer (or online at OneDrive) by scrolling to the Choose a File section and clicking Browse Files. Then navigate to and select the picture you want. *Or...*

5. You can take a picture with your computer's webcam to use for your account picture. In the Take a Photo section, click Open Camera and follow the onscreen directions from there.

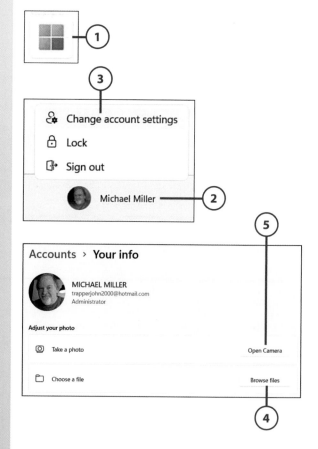

Configuring Other Windows Settings

You can configure many other Windows system settings. In most cases, the default settings work fine and you don't need to change a thing. However, you *can* change these settings if you want to or need to.

Configure Settings from the Settings App

You configure most Windows settings from the Settings app, which consists of a series of tabs, accessible from the left side of the window, that present different types of settings.

1. Click or tap the Start button to display the Start menu.

2. Click or tap Settings to open the Settings app.

3. To search for a specific setting, type your query into the Find a Setting box and click the Search (magnifying glass) icon. *Or...*

4. Click or tap a category in the left panel to display settings of that type.

5. Click or tap to select the type of settings you want to configure.

6. Configure the necessary options.

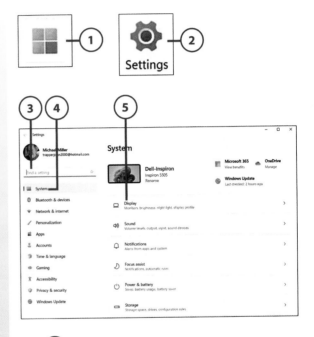

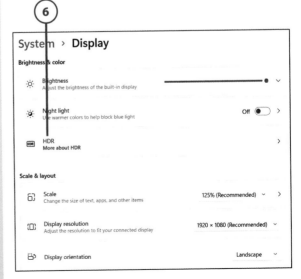

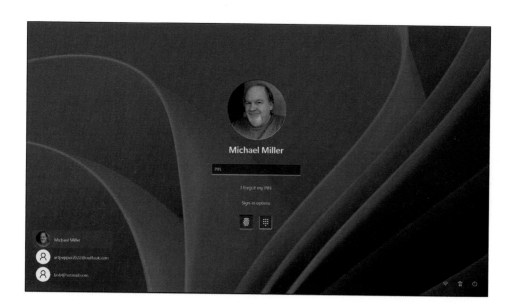

In this chapter, you find out how to share your computer with other users.

→ Understanding User Accounts
→ Adding New Users
→ Signing In and Switching Users

Working with Different Users

Chances are you're not the only person using your computer; it's likely that you'll be sharing your PC with your spouse or partner and maybe even your children and grandchildren. Fortunately, you can configure Windows so that different people using your computer sign on with custom settings—and have access to their own personal files.

Understanding User Accounts

The best way for multiple people to use a single computer is to assign each person their own password-protected *user account*. For a given person to use the PC and access their own programs and files, they have to sign in to the computer with their personal password. If a person doesn't have an account or the proper password, they can't use the computer.

Windows lets you create two different types of user accounts—online and local. The default is the online account, which comes with some unique benefits.

An online account is linked to a new or existing Microsoft account. When you use a Microsoft account on your computer, Windows displays information from other Microsoft sites you use. For example, Windows displays the latest weather conditions in the Weather app, the latest news headlines in the News app, and the latest stock quotes in the Stock app—all based on settings you make when you configure your Microsoft account. Local accounts cannot access this personalized data.

For these reasons, I recommend you create a (or use an existing) Microsoft account for any new user you add to your computer.

>>>*Go Further*

DIFFERENT ACCOUNTS FOR DIFFERENT USERS

When you get your PC set up just the way you like, you may be hesitant to let anybody else use it. This goes double for your children and grandchildren; you love 'em, but don't want them to mess up your computer with their games and tweeting and whatnot.

This is where creating separate user accounts has value. Create a user account for each user of your PC—for you, your spouse, and each of your kids and grandkids—and then make everybody sign in under their personal accounts. The other family members can personalize their accounts however they want, and there's nothing they can change in your account. The next time you sign in to your user account, everything should look just the way you left it—no matter who used your computer in the meantime.

Adding New Users

You create or link to one user account when you first launch Windows on your new PC. At any time, you can create additional user accounts for other people using your computer.

Add a User with an Existing Microsoft Account

By default, Windows tries to use an existing Microsoft account to create your new Windows user account. So, if you have an Outlook.com, OneDrive, Skype, Xbox Live, or other Microsoft account, you can use that account to sign in to Windows on your computer.

(1) Click or tap the Start button to display the Start menu.

(2) Click or tap Settings to open the Settings app.

(3) Click or tap Accounts to display the Accounts page.

(4) Click or tap Family & Other Users.

(5) In the Add a Family Member section, click or tap Add Account to display the Microsoft Account window.

(6) Enter the person's email address.

(7) Click or tap Next.

Settings

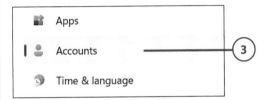

- Apps
- Accounts   **(3)**
- Time & language

Family & other users
Device access, work or school users, kiosk assigned access

(4)   **(5)**

Add a family member   Add account

(6)

Microsoft account   ✕

Add someone

Enter their email address

No Microsoft account? Create one for a child

Cancel   Next

(7)

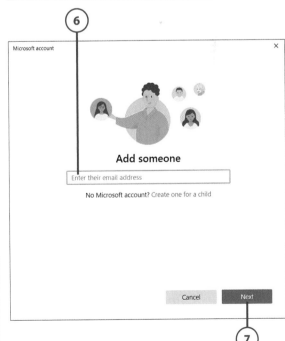

8. Select either Organizer (to let the new user edit family and safety settings) or Member (for younger users who you don't want to allow to edit these settings).

9. Click Invite. The person you entered is immediately added as a new user to your computer. They will receive an invitation via email to join your family group.

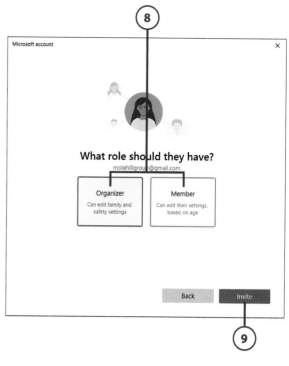

Create a New User Account

If you're adding a user to your computer who does not have an existing Microsoft account, you need to create a new Microsoft account for that person. This gives that individual a new Outlook.com email address, which they can use to log into Windows on your PC.

1. Click or tap the Start button to display the Start menu.

2. Click or tap Settings to open the Settings app.

3. Click or tap Accounts to display the Accounts page.

4. Click or tap Family & Other Users.

5. Scroll to the Other Users section and click or tap Add Account to display the Microsoft Account window.

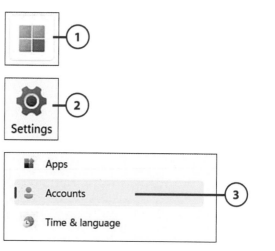

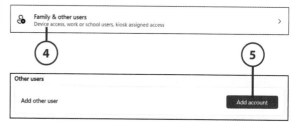

(6) Click or tap I Don't Have This Person's Sign-In Information.

(7) Click or tap Get a New Email Address to display the Create Account page.

(8) Click to pull down the domain list and make sure outlook.com is selected.

(9) Enter the desired email username into the New Email field and then click Next. (You might have to try several names to get one that isn't already taken.)

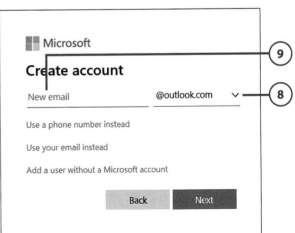

(10) Enter the desired password into the Create a Password box.

(11) If you don't want this person to receive Microsoft junk mail, uncheck the I Would Like Information, Tips, and Offers About Microsoft Products and Services box.

(12) Click Next.

Passwords

Passwords must be at least eight characters long and contain at least two of the following: uppercase letters, lowercase letters, numbers, and symbols.

(13) Enter the person's first and last name.

(14) Click Next.

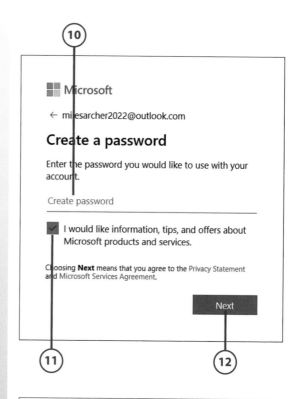

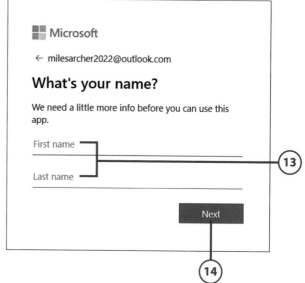

15. Select the region where this person lives.

16. Enter this person's birthdate.

17. Click Next. The new account is created.

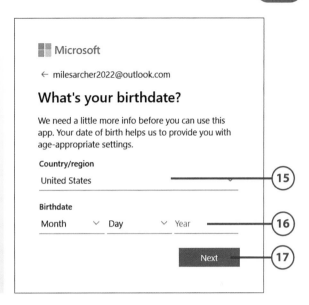

>>>Go Further
CHILD ACCOUNTS

If you're setting up an account on your computer for a younger family member, you might want to make that account a special *child account*. You see this option when you click Add Account in the Add a Family Member section.

The big difference between a child account and a regular account is that Windows enables Family Safety Monitoring for the child account. With Family Safety Monitoring, you can turn on web filtering (to block access to undesirable websites), limit when younger children or grandchildren can use the PC and what websites they can visit, set limits on games and Windows Store app purchases, and monitor the kids' PC activity.

To do this, open the Settings app, select Privacy & Security, click Windows Security, and click Family Options. This opens the Windows Security window with the Family Options page selected. Click View Family Settings.

This opens a web browser and takes you to your Microsoft account online; from there, you can turn on and off Family Safety and activity reporting, as well as configure web filtering, time limits, and Microsoft Store, game, and app restrictions for any account.

It's all about making Windows—and your Windows computer—safer for younger users. The younger members of your family will be the better for it.

Signing In and Switching Users

If other people are using your computer, they'll want to sign in with their own accounts. Fortunately, it's relatively easy to sign in and out of different accounts and to switch users.

Set Sign-In Options

There are several different types of security you can employ when users are signing into your PC. A user can sign in with any of the following methods:

- Traditional alphanumeric password
- Windows Hello PIN (personal identification number)
- Windows Hello Face facial recognition (on compatible PCs)
- Windows Hello Fingerprint (on compatible PCs)
- Security Key (on a USB drive-like device)
- Picture password (requires you to sketch a portion of a picture onscreen)

Each user can select which sign-in option they want to use.

1. Click or tap the Start button to display the Start menu.

2. Click or tap Settings to open the Settings app.

3. Click or tap Accounts to display the Accounts page.

4. Click or tap Sign-In Options.

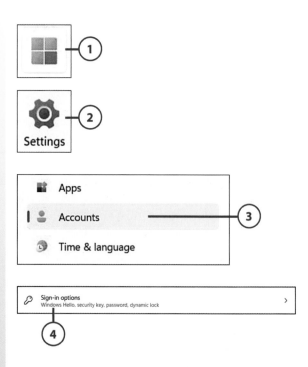

5 Click the sign-in method you want to use and then follow the onscreen instructions to implement that method. (For example, if you choose the Password option, you need to enter the desired password.)

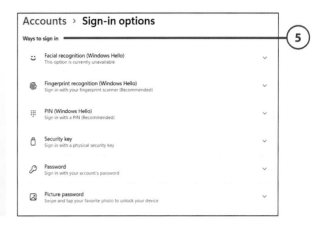

Accounts > **Sign-in options**

Ways to sign in   **5**

☺   Facial recognition (Windows Hello)   ⌄
     This option is currently unavailable

🔘   Fingerprint recognition (Windows Hello)   ⌄
     Sign in with your fingerprint scanner (Recommended)

⠿   PIN (Windows Hello)   ⌄
     Sign in with a PIN (Recommended)

🔒   Security key   ⌄
     Sign in with a physical security key

🔑   Password   ⌄
     Sign in with your account's password

🖼   Picture password   ⌄
     Swipe and tap your favorite photo to unlock your device

Passwords and PINs

You may be used to using a password to sign into Windows, but Microsoft is now encouraging using other more secure methods instead. If your PC is compatible, fingerprint or facial recognition are the most secure and easiest methods available. Absent those, Microsoft recommends using a five-digit PIN instead of a password because the PIN is more difficult for hackers to guess.

>>>Go Further
OTHER SIGN-IN OPTIONS

There are a few other options on the Sign-In Options page that can affect your computer's security and ease of use. These include the following:

- Only allow Windows Hello sign-in on this device (removes the options for signing in with a password or picture password)
- Should Windows require you to sign in again if you've been away from your computer
- Dynamically lock your computer when you're away
- Automatically save restartable apps and restart them when you sign back on
- Show account details on the sign-in screen
- Use sign-in information to automatically finish setting up after a system update

The default settings are generally the best for all these settings—unless, of course, you prefer to use a password rather than a PIN; then you want to deselect the Windows Hello requirement option.

Sign In with Multiple Users

If you have more than one user assigned to Windows, the sign-in process is slightly different when you start up your computer.

1. Power up your computer.

2. When the Windows lock screen appears, press any key on your keyboard, tap or touch the touchpad, or click the mouse to display the sign-in screen. All users of this computer are listed here.

3. Select your username.

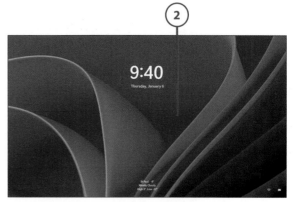

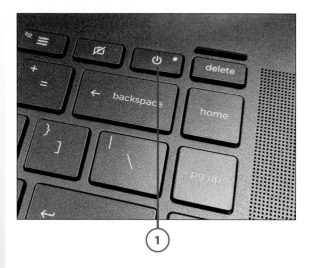

4 Enter your PIN or password (or use another sign-in option) and then press the Enter key to display your personal desktop.

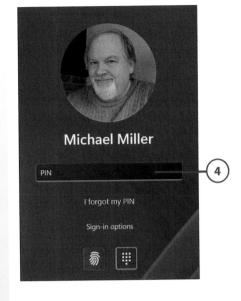

Switch Users

You can also change users without restarting your PC.

1 Click or tap the Start button to display the Start menu.

2 Click or tap your name or picture at the bottom of the Start menu to display a list of other users.

3 Click the desired user's name.

4 When prompted, enter the new user's sign-in information and then press Enter.

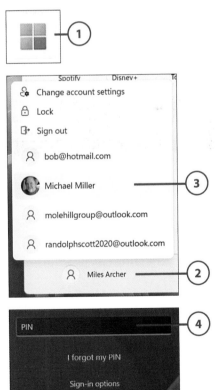

In this chapter, you discover how to use software applications in Windows 11.

8

Using Apps and Programs

You can run two types of programs in Windows 11. First are those
Windows apps you get from the Microsoft Store, designed to work
specifically in the Windows 11 environment. Then there are traditional
software programs that run on the Windows desktop. The two types of
programs are subtly different.

Finding and Launching Apps in Windows

All the applications you have installed on your PC, both traditional
software programs and more modern Windows apps, are accessible
from the Windows Start menu. Opening an app is as easy as clicking or
tapping that app's icon.

Display All Apps

By default, when you open the Start menu, you see those apps you've pinned there. Viewing all the apps installed on your computer takes an extra step.

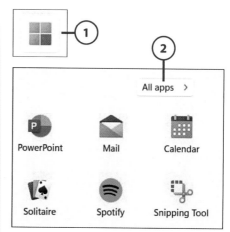

1. Click or tap the Start button to display the Start menu. By default, only your pinned apps are visible.

2. Click or tap All Apps to view an alphabetical list of all your apps.

3. Scroll down the list to view all the apps installed on your computer.

Pinning Apps

To learn how to pin your favorite apps to the Start menu, turn to Chapter 6, "Personalizing Windows."

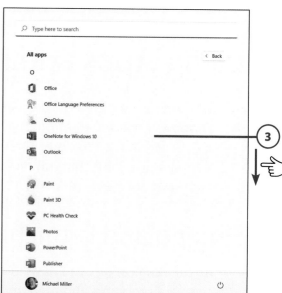

Search for Apps

If you have a lot of apps installed on your PC, you may have to scroll quite a bit to find a particular app on the Start menu. It may be easier to search for the app you want.

1. Click or tap the Search icon on the taskbar to display the Search pane.

2. Enter the name of the app into the Type Here to Search field. As you type, Windows displays a list of apps, files, and settings that match your query.

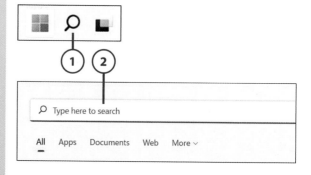

(3) Limit the results to just apps by clicking or tapping the Apps tab.

(4) Click or tap an app name to open it or select open from the Details pane on the right.

Searching for Files

You can also use the Search pane to search for specific files stored on your computer, which is great when you have no idea where a particular file is stored. Just select the Documents tab instead of the Apps tab and search as normal.

Search from the Start Menu

You can also search from the Start menu. Just click or tap within the Search field at the top of the Start menu to display the Search pane.

Open an App

How you open an app depends on where you are in Windows.

(1) From the Windows Start menu, click or tap the icon for the app. *Or...*

(2) If an app is pinned to the taskbar, click or tap that app's icon there. *Or...*

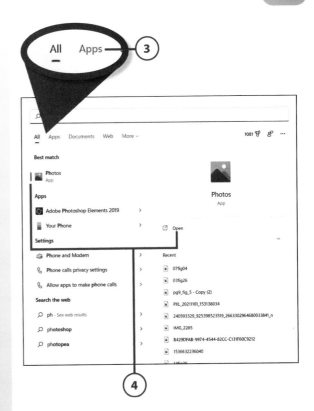

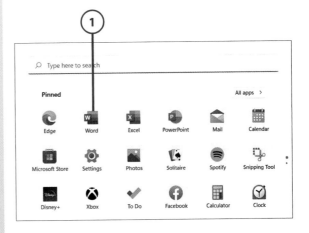

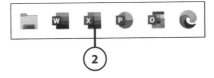

(3) If there's a shortcut for the app on the Windows desktop, *double-click* or double-tap the shortcut to open the app.

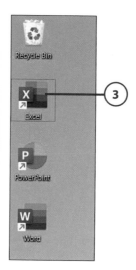

Working with Apps

Every Windows app or software program opens in an individual window on the Windows desktop. You can easily resize and rearrange all the open windows on the desktop.

Scroll Through a Window

Many programs, documents, and web pages are longer than the containing window is tall. To read the full page or document, you need to scroll through the window.

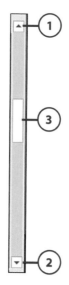

(1) Click or tap the up arrow on the window's scroll bar to scroll up one line at a time.

(2) Click or tap the down arrow on the window's scroll bar to scroll down one line at a time.

(3) Click and drag the scroll box (slider) to scroll up or down in a smooth motion.

Other Ways to Scroll

You can also scroll up or down a window by pressing the Page Up (PgUp) and Page Down (PgDn) keys on your keyboard. In addition, if your mouse has a scroll wheel, you can use it to scroll through a window.

Maximize, Minimize, and Close a Window

After you've opened a window, you can maximize it to display full screen. You can also minimize it so that it disappears from the desktop and resides as a button on the Windows taskbar, and you can close it completely.

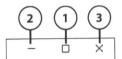

1. Maximize the window by clicking or tapping the Maximize button in the top-right corner. When the window is maximized, the Maximize button turns into a Restore Down button; click or tap this button to return the window to its original size.

2. Minimize the window by clicking or tapping the Minimize button in the top-right corner. The window shrinks to an icon on the taskbar; to restore the window to its original size, click or tap the window's icon on the taskbar.

3. Close the window completely (and shut down the program or document inside) by clicking Close (the X) at the top-right corner.

Resize a Window

You can resize any individual window to fill the entire screen or just a part of the screen.

(1) Use your mouse or touchpad to put your cursor over any side or corner of the window; the cursor should turn into a double-sided arrow. Click and hold the mouse button or press on the touchpad.

(2) Keep the mouse button pressed or keep your finger on the touchpad and then drag the edge of the window to a new position.

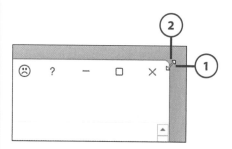

Snap a Window

Any open window can be "snapped" to the left or right side of the desktop so it shares the screen with another app. Windows 11 offers a variety of snapping options, including one that lets you display four different windows onscreen at the same time!

(1) Within the first app, mouse over the Maximize button.

(2) This displays all available snap layouts. Click or tap the position you want this app to be within the given layout.

(3) The current app is sized and positioned according to the selected layout. Thumbnails of all other open apps also appear on the desktop within the next position in the layout. Click or tap the app you want to appear in this position.

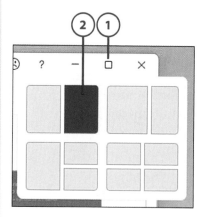

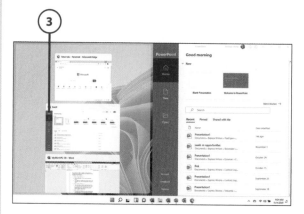

4 For layouts with three or more apps, the remaining open apps appear in the next open layout position. Click or tap the app you want to appear in this position.

5 For the four-app layout, click or tap the open app you want to appear in the remaining position.

6 To display a window full screen, click or tap the Maximize button for that app.

Move a Window

To move a window to another position on the desktop, click and drag the window's title area.

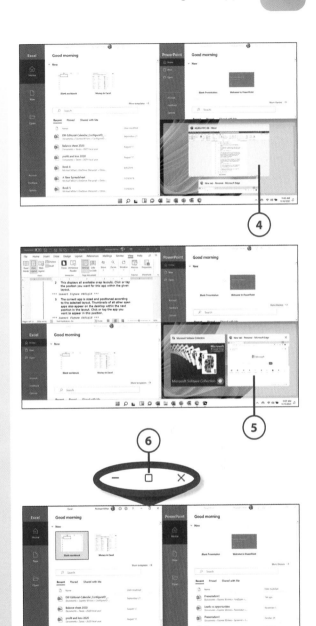

Use Pull-Down Menus, Toolbars, and Ribbons

Most traditional software programs have similar onscreen elements—menus, toolbars, ribbons, and such. Once you learn how to use one program, the others should be quite familiar.

(1) Many software programs use a set of pull-down *menus* to store all the commands and operations you can perform. The menus are aligned across the top of the window, just below the title bar, in what is called a *menu bar*. Click or tap the menu's name to pull down the menu, and then click a menu item to select it.

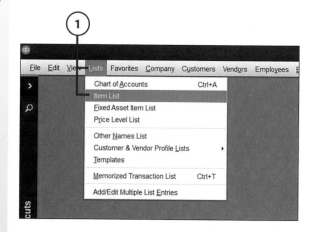

(2) Some older software programs put the most frequently used operations on one or more *toolbars*, which are usually located just below the menu bar. A toolbar looks like a row of buttons, each with a small picture (called an *icon*) and maybe a bit of text. Click or tap a button on the toolbar to select that operation.

(3) Many newer software programs, including Microsoft Office, use a ribbon interface that contains the most frequently used operations. A *ribbon* is typically located at the top of the window, beneath the title bar (and sometimes the menu bar). Ribbons often consist of multiple tabs; select a tab to see buttons and controls for related operations. Click or tap a button on the ribbon to select that operation.

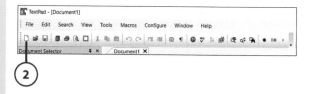

Display or Hide

If the full ribbon isn't visible, click the down arrow at the far-right side of the tabs. To minimize the ribbon and its buttons, click the up arrow at the far-right side of the ribbon.

Switch Between Open Windows

After you've launched a few programs, you can easily switch between one open program and another.

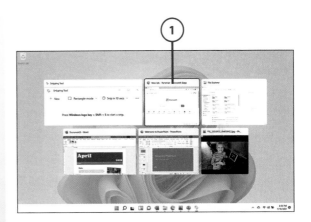

1. Press Alt+Tab to display thumbnails of all open windows. Keep pressing Tab to cycle through the open apps. Release the keys to switch to the selected window.

2. Alternatively, click or tap the Task View button on the taskbar. This displays thumbnails of all open windows. Click or tap the window you want to switch to.

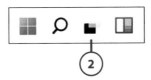

3. When a program or document is open, an icon for that item (with a line underneath) appears in the Windows taskbar. Mouse over that icon to view a thumbnail preview of all open documents for that application. To switch to an open document, click or tap the thumbnail for that document.

Multiple Documents

If multiple documents or pages for an application are open, multiple thumbnails appear when you hover over that application's icon in the taskbar.

Work with Multiple Desktops in Task View

Windows 11 enables you to open multiple apps and save that combination of apps as a unique virtual desktop. For example, you might create one desktop with all your work apps and another with your social media apps and then switch between the two desktops.

1. On the taskbar, click or tap the Task View button to open Task View.

2. Click or tap the New Desktop tile.

3. Open any apps that you want to appear on this desktop by opening the Start menu and selecting apps you want to open.

4. Click the Task View button to cycle among your desktops.

5. Click the thumbnail for the desktop you want to open.

Naming and Deleting Desktops

To name or rename a desktop, click the Task View button, right-click the desktop you want to rename, select Rename, and then enter a new name. To delete a desktop, open Task View, mouse over the desktop you want to delete, and then click the Close (X) button

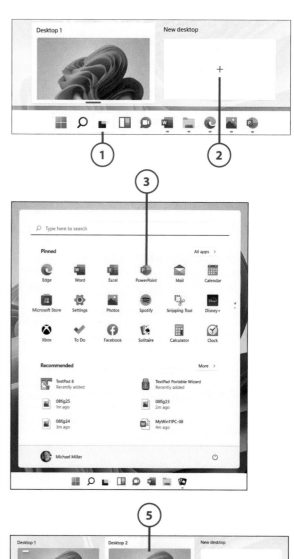

Shopping for Apps and Programs

There are many more apps and programs available than those included with Windows 11. You can install both modern Windows apps and traditional software programs on your Windows 11 PC.

Preinstalled Apps

Windows 11 comes with several useful apps preinstalled. These include Calculator, Calendar, Clock, Microsoft News, Notepad, and Weather. You can open all of these from the Start menu.

Find and Install Windows Apps

The first place to find new apps is in the online Microsoft Store, which has been totally revamped for Windows 11. The Microsoft Store includes a mix of newer-style Windows apps and traditional desktop apps.

You access the Microsoft Store from the Start menu, as if it were another app. Click or tap the Store icon to start the app.

Pricing

Whereas a traditional computer software program can cost hundreds of dollars, many of the apps in the Windows Store cost $20 or less—and some are available for free. (Note that some of the free apps require in-app purchases to activate additional functionality.)

(**1**) The Microsoft Store launches with featured apps and games at the top of the home page. Scroll down to view different types of apps and games.

(**2**) Click the Apps tab to view only apps. *Or…*

(**3**) Click the Gaming tab to view only games. *Or…*

(**4**) Click the Movies & TV tab to view movies, TV programming, and other videos for purchasing. *Or…*

(**5**) To search for a specific app, type the app's name into the Search box at the top of the window.

(**6**) To view more information about an app, click the app to display the app's information page.

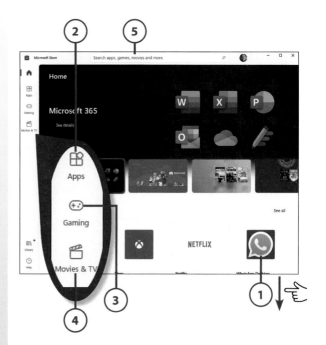

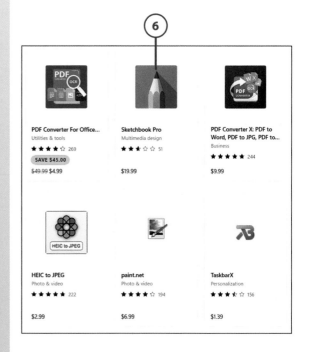

7 Scroll down to read a description of the app as well as system requirements and reviews that other users have left.

8 Click the price button to purchase the app; when prompted to confirm your purchase, click the Confirm button.

9 If the app is free, click the Install button to download and install it on your computer.

First-Time Purchase

The first time you purchase an app from the Microsoft Store, you're prompted to set up a payment method. Do so, and Microsoft remembers this information for future purchases.

Find and Install Traditional Software Programs

The new Microsoft Store features both Windows apps and traditional desktop software programs. You can also find traditional desktop software in-store or online from many consumer electronics stores, office stores, computer stores, and mass merchants (such as Best Buy, Target, and Walmart).

Some software programs you purchase at retail stores still come with a CD or a DVD disc; these disks typically come with built-in installation utilities. Just insert the program's disc into your computer's CD/DVD drive (if it has one). The installation utility should run automatically.

Other software programs don't include any media in the box because the manufacturers recognize that most newer PCs don't come with CD/DVD drives. Instead, you're given instructions for downloading the app from the manufacturer's website and a code to activate the program you paid for.

In addition, many software publishers make their products available via download from the Internet. When you download a program from a major software publisher, the process is generally easy to follow. Just click the "buy" button and follow the onscreen instructions.

It's Not All Good

Download from Legitimate Sites Only

Limit your software downloads to reputable download sites and software publisher sites. Make sure the website has an address that begins with https:// (not just http://) and features a lock icon in the browser address box. This ensures that you're working over a secure Internet connection.

One serious issue with programs downloaded from unofficial sites is that they might contain computer viruses or spyware (called *malicious software*, or *malware*), which can damage your computer. It's a lot safer to deal with official sites that regularly scan their offerings to ensure that they're not infected in this manner.

Even legitimate download sites might contain confusing advertisements that look like download links or buttons. Do not click these misleading links because you might end up downloading unwanted malware. Make sure you find the correct download link or button and click it only.

And, when you're installing the software, be careful what options you click. Some otherwise-legitimate programs attempt to install other unwanted software during their installation processes. Read every onscreen message carefully, and only click to approve those items you want to install.

In this chapter, you discover how to use the accessibility functions built in to Windows 11 and other ways to make Windows easier to use if you have vision or mobility issues.

→ Using Accessibility Functions in Windows 11
→ Using Alternative Input Devices

9

Making Windows Easier to Use

If you have 20/20 vision, perfect hearing, and the Samson-like grip of a circus strongman, good for you. For some of us, however, the default display settings of most computers, particularly laptop models with smaller screens and cramped keyboards, can affect our ability to use our PCs.

Fortunately, Microsoft offers a number of accessibility features that can make Windows—and your new PC—a little easier to use. Let's take a look.

Using Accessibility Functions in Windows 11

The accessibility features in Windows 11 are designed to make your computer easier and more comfortable to use, especially if you have vision, hearing, or dexterity issues. Microsoft offers several useful accessibility functions, including the capability to enlarge text on the screen, change the contrast to make text more readable, and read the screen to aid those with vision problems.

Access Accessibility Features

The easiest way to get to the accessibility settings is through the Settings app.

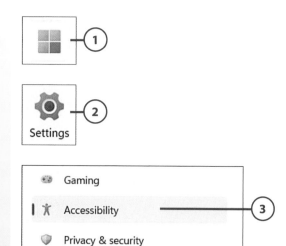

(1) Click or tap the Start button to display the Start menu.

(2) Click or tap Settings to open the Settings app.

(3) Click or tap Accessibility to display the Accessibility screen.

Enlarge the Screen

If you're having trouble reading what's onscreen because the text is too small, you can turn on the Magnifier tool. The Magnifier does just what the name implies—it magnifies an area of the screen to make it larger and easier to read.

(1) From the Vision section of the Accessibility screen, click or tap Magnifier.

(2) Click or tap "on" the Magnifier control—that is, move the Magnifier slider to the On position.

Keyboard Shortcut

You can also enable Magnifier by pressing the Windows key and the plus (+) key on the numeric keyboard.

Invert Colors

To display the magnified screen with inverted colors (white text against a black background), scroll down the Magnifier page and select the Invert Colors option.

③ The screen enlarges to 200% of its original size. Navigate around the screen by moving your mouse to the edge of it. (For example, to move the screen to the right, move your mouse to the right edge of the screen.)

④ Click or tap the + (plus) button on the Magnifier control to enlarge the screen further.

⑤ Click or tap the – (minus) button on the Magnifier control to reduce the size of the screen. (Alternatively, press Windows + – [the minus key] on your keyboard.)

⑥ Click or tap the X to turn off the Magnifier. (Alternatively, press Windows + Esc on your keyboard.)

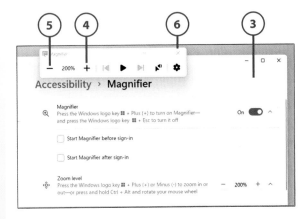

>>>Go Further

MAKING WEB PAGES EASIER TO READ

If you're like me, you spend a lot of time reading articles and other content on the Web. Unfortunately, many web pages are difficult to read, cluttered with unnecessary ads and images and with text that's just a little too small.

There are a few things you can do to make web pages easier to read. The first is to increase the size of the onscreen text. In the Microsoft Edge browser, you do this by clicking or tapping the Settings and More (three-dot) button, and then clicking or tapping the + zoom control. If you're using the Google Chrome browser, click or tap Customize and Control in the top-right corner to access the zoom control.

There's an even better approach in Microsoft Edge—a new feature called the Immersive Reader. With the Immersive Reader, Edge removes all the unnecessary ads and graphic elements; it displays only the text and accompanying pictures for the current web page. The Immersive Reader also increases the text size, so all around you get a much better reading experience. For those with vision problems, the Immersive Reader is a godsend. Check it out by going to your favorite web page and then clicking or tapping the Immersive Reader button in the browser's Address bar (or pressing F9 on your keyboard). It really works!

Learn more about the Immersive Reader, and web browsing in general, in Chapter 11, "Browsing and Searching the Web."

Use Color Filters

If you experience color blindness, it may be easier to see elements of the Windows desktop in grayscale or other color shades rather than their original colors.

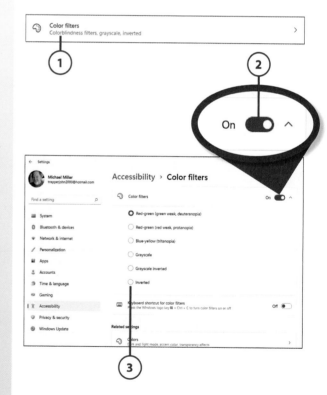

1. From the Vision section of the Accessibility screen, click or tap Color Filters.

2. Click or tap "on" the Turn On Color Filters switch.

3. Select a filter from the Choose a Filter list—Red-green (green weak, deuteranopia), Red-green (red weak, protanopia), Blue-yellow (tritanopia), Grayscale, Grayscale Inverted, or Inverted.

Improve Onscreen Contrast

Some people find it easier to view onscreen text if there's more contrast between the text and the background. To that end, Windows 11 lets you switch to a high-contrast mode that displays lighter text on a dark background instead of the normal black-on-white theme.

1. From the Accessibility screen, click or tap Contrast Themes.

2. Click or tap to pull down the Contrast Themes list.

3. Select a theme.

4. Click or tap the Apply button.

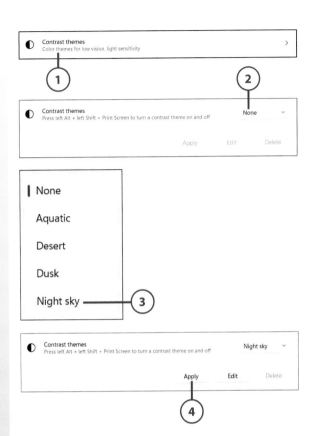

Make the Mouse Pointer Easier to See

Another issue that many users have is seeing the mouse pointer onscreen. The default mouse pointer in Windows can be a little small and difficult to locate on a busy desktop; you can change the size and color of the pointer to make it easier to see. You can choose from Regular, Large, and Extra Large size settings, as well as from White, Black, and Inverting color settings.

1. From the Vision section of the Accessibility screen, click or tap Mouse Pointer and Touch.

(2) In the Mouse Pointer Style section, select a different pointer—ideally, one that's easier to see onscreen.

(3) In the Size section, select a larger pointer size.

Touchscreen Considerations

If your computer has a touchscreen display, you can configure Windows to show an onscreen circle where you touch the screen. From the Mouse Pointer and Touch screen, click or tap "on" the Touch Indicator switch. You can then opt to make the onscreen circle larger and darker if you want.

Make the Text Cursor Easier to See

If you have trouble seeing the mouse pointer, you may also have trouble seeing the text cursor when you're editing documents onscreen. Windows 11 lets you display colored text cursor indicators above and below the normal text color—and change the thickness of the cursor.

(1) From the Vision section of the Accessibility screen, click or tap Text Cursor.

(2) Click or tap "on" the Text Cursor Indicator switch to turn on color shapes above and below the normal text cursor.

(3) Drag the Size slider to select a larger or smaller size for the text cursor indicator.

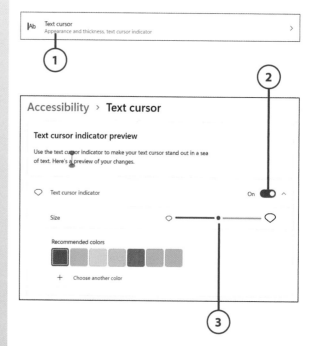

4. In the Recommended Colors section, select a different color for the text cursor indicator.

5. Scroll to the Text Cursor Thickness section and adjust the Text Cursor Thickness slider to make the cursor thinner or thicker.

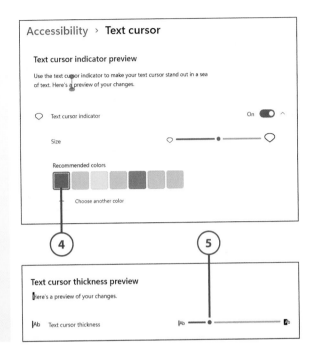

>>>Go Further

CONNECTING A LARGER SCREEN

The easiest solution if you're having trouble seeing what's onscreen is to make the screen bigger—literally. This means connecting a larger computer monitor to your computer. Many people find that monitors sized 22" and larger are a lot easier to see than the standard 15" screens that are common on today's laptops. See Chapter 5, "Connecting Printers and Other Peripherals," for detailed instructions.

Read Text Aloud with Narrator

If your eyesight is really bad, even making the onscreen text and cursor super large won't help. To that end, Windows offers the Narrator utility, which speaks to you through your PC's speakers. When you press a key, Narrator tells you the name of that key. When you mouse over an item onscreen, Narrator tells you what it is. Narrator helps you operate your PC without having to see what's onscreen.

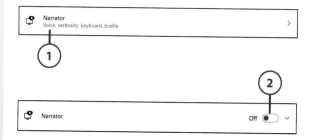

1. Scroll to the bottom of the Vision section of the Accessibility screen and click or tap Narrator.

2. Click or tap "on" the Narrator switch.

Keyboard Shortcut
You can also enable Narrator by pressing Windows+Ctrl+Enter on your keyboard.

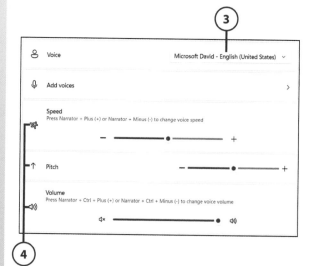

3. Pull down the Voice list and select from either David, Zira, or Mark.

4. Use the Speed, Pitch, and Volume controls to adjust how the voice sounds to you.

Narrator Home
Click or tap Narrator Home at the top of the Narrator page to explore more Narrator options in the Narrator Home app.

>>>*Go Further*
MORE ACCESSIBILITY OPTIONS

I've covered the major accessibility functions here, but there are more where these came from. If you're having difficulty seeing items onscreen or operating Windows, you should explore all the options present on the Accessibility settings screen. You can, for example, enable visual notifications for Windows sounds, opt to use the numeric keypad to move the mouse around the screen, reconfigure scroll bars and other visual effects, and choose to activate a window by hovering over it with your mouse.

Use the On-Screen Keyboard

If you find that pressing the keys on your computer keyboard with your fingers is becoming too difficult, especially on a laptop PC with smaller keys, you might want to use the Windows On-Screen Keyboard. This is a virtual keyboard, displayed on your computer screen, that you can operate with your mouse instead of your fingers (or with your fingers, if you have a tablet PC without a traditional keyboard).

1. From the Accessibility screen, scroll to the Interaction section and click or tap Keyboard.

2. Click or tap "on" the On-Screen Keyboard switch. The On-Screen Keyboard displays.

Keyboard Shortcut
You can also display or hide the On-Screen Keyboard by pressing Win+Ctrl+O on the normal computer keyboard.

3. To "press" a key, click it with your mouse—or, on a touchscreen display, tap it with your finger.

4. To close the On-Screen Keyboard, click or tap the X in the top-right corner.

>>>Go Further

QUICK ACCESS TO ACCESSIBILITY OPTIONS

You can access the accessibility options from the Settings app, as discussed, or from the Quick Access panel. Click or tap the middle of the notification area of the taskbar to display the Quick Access panel; then click or tap Accessibility. This lets you quickly turn on or off Magnifier, Color Filters, Narrator, Mono Audio, and Sticky Keys.

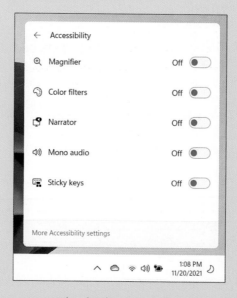

(Mono Audio feeds the same sound to both stereo speakers on your PC to make it easier to hear some things. Sticky Keys lets you enter multiple-key shortcuts by pressing one key at a time instead of pressing them simultaneously.)

Using Alternative Input Devices

Some of us might lose fine mobility in our hands and fingers, whether due to arthritis or some other condition. This might make it difficult to use the small touchpad on most laptop PCs or to type on normal-sized keyboard keys.

The solution for this problem is to attach different input devices. You can easily connect an external mouse to a laptop PC that then replaces the built-in touchpad or attach a keyboard with larger keys for easier use.

Replace the Touchpad

Touchpads are convenient pointing devices for laptop PC users, but they can be difficult to use, especially if you have difficulty moving or holding your hand and fingers steady. The solution is to attach an external pointing device to a USB port on your laptop PC; when you do this, you use the (hopefully easier-to-use) external device instead of the built-in touchpad. There are two primary types of devices to choose from.

The first one is a simple external mouse that connects wirelessly to your computer via either USB cable or wirelessly via Bluetooth. Many people find that using a mouse is easier than trying to tap precise movements on a laptop's touchpad. If you're using a wireless mouse, connect the mouse's USB receiver into any open USB port on your computer, turn on the external mouse, and start using it. In most instances, no additional setup is required.

In lieu of a touchpad or mouse, some users prefer an even larger trackball controller. This type of controller is typically used by people who play computer games, but it's also a terrific option for those with mobility issues. Use the large roller ball on top to move the cursor around the screen.

Attach a Different Keyboard

Some laptop PC keyboards are a little smaller than the keyboard on a typical desktop PC, and they may have flatter keys that don't respond much to your touch. You can remedy this situation by attaching a full-size external keyboard, either via USB cable or wirelessly via Bluetooth. Most external keyboards are easier to use and more ergonomic than the smaller keyboards found on most laptop computers.

Even the keys on a standard-sized keyboard might not be big enough if you have mobility issues, however. Several companies make keyboards with enlarged keys that are both easier to see and easier to use. Look for these online.

In this chapter, you find out how to connect to the Internet from your home network or at a public Wi-Fi hotspot.

→ Connecting to the Internet—and Your Home Network
→ Connecting to the Internet at a Public Wi-Fi Hotspot

10

Connecting to the Internet—at Home or Away

Much of what you will use your new computer for is on the Internet. The Internet is a source of information, a conduit for shopping, banking, and other useful activities, a place to play games, and a tool for communicating with friends and family.

To get full use out of your new PC, you need to connect it to the Internet. You can connect to the Internet at home or away. All you need is access to a home network or, outside of your house, a Wi-Fi hotspot.

Connecting to the Internet—and Your Home Network

To get Internet in your home, you need to contract with an *Internet service provider* (ISP). You can typically get Internet service from your cable company or from your phone company. Prices vary depending on the speed of the Internet service provided, and whether you bundle it with other services (such as cable or phone).

Your ISP should set you up with a broadband *modem* that connects to the incoming cable or phone line. The modem takes the digital signals coming through the incoming line and converts them into a format that your computer can use.

In most cases, you connect your broadband modem to a *wireless router*. A router is a device that takes a single Internet signal and routes it to multiple devices; when you set up your router, following the manufacturer's instructions, you create a *wireless home network*. You connect your computer (as well as your smartphone, tablet, and other wireless devices) wirelessly to your router via a technology called *Wi-Fi*.

Wireless Gateway

Instead of a separate modem and router, many ISPs supply a *wireless gateway* that combines those two devices in a single unit. This makes for easier connection and setup—you have one less cable to run and one less device to configure.

Connect to Your Home Network

After your modem and wireless router (or combined gateway) are set up, you've created a wireless home network. You can then connect your PC to your wireless router—and access the Internet.

Switching Wi-Fi On and Off

Your computer's Wi-Fi needs to be switched on to connect to the Internet. It's typically switched on by default, but you can switch it on and off manually if you want. Click or tap the middle of the notification area of the taskbar to display the Quick Settings panel and click the left side of the Wi-Fi button. The button is blue when Wi-Fi is on.

1 Click or tap the middle of the notification area of the taskbar to display the Quick Settings panel.

2 Click or tap the right arrow on the Manage Wi-Fi Connections button. You see a list of available wireless networks.

Secured Versus Nonsecured Networks

A wireless network that is secured by a password, such as your home network, is displayed with a lock icon next to its name. A network that doesn't require a password, such as a public Wi-Fi hotspot, doesn't display the lock icon.

3 Click or tap to select your wireless network. This expands that network's section.

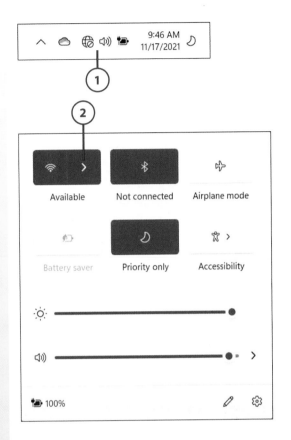

4 Check the Connect Automatically option to connect automatically to this network in the future.

5 Click or tap the Connect button. This expands this section.

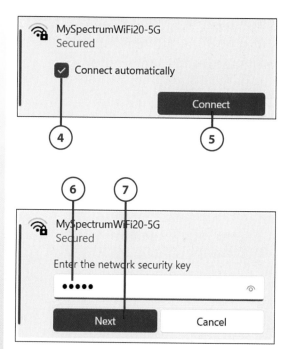

Connect Automatically

When you're connecting to your home network, it's a good idea to enable the Connect Automatically feature. This lets your computer connect to your network without additional prompting or inter- action on your part.

6 When prompted, enter the pass- word (called the *network secu- rity key*) for your network. This password is provided with your network router, either printed on the router itself or in the router's instructions, or it may be manu- ally assigned.

7 Click or tap Next. Your computer is now connected to the network.

One-Button Connect

If the wireless router on your network supports "one-button wireless setup" (based on the Wi-Fi Protected Setup, or WPS, technology), you might be prompted to press the "connect" or WPS button on the router to connect. This is much faster than going through the process outlined here.

Access Other Computers on Your Network

Once your computer is connected to your home network, you can access the content on other computers on the same network.

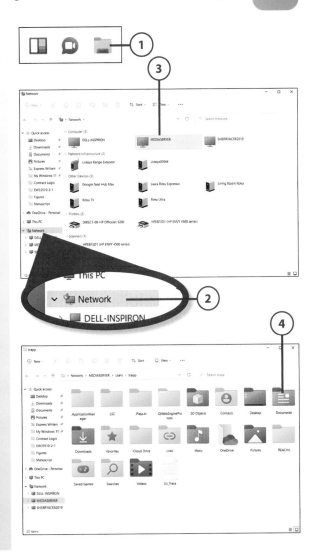

(1) Click or tap the File Explorer icon on the taskbar or Start menu to open File Explorer.

(2) Click Network in the navigation pane. This displays all the computers and devices connected to your network.

(3) Double-click the computer you want to access.

(4) Windows displays the shared folders on the selected computer. Double-click a folder to view that folder's content.

Sharing Content

You can access content only on other computers that have been configured as shareable—that is, the computer owner has enabled sharing for that particular folder or type of content.

Connecting to the Internet at a Public Wi-Fi Hotspot

The nice thing about the Internet is that it's virtually everywhere. This means you can connect to the Internet even when you're away from home. All you need to do is find a wireless connection, called a *Wi-Fi hotspot*. Fortunately, most coffeehouses, libraries, hotels, fast-food restaurants, and public spaces offer Wi-Fi hotspots—often for free.

Connect to a Wi-Fi Hotspot

When you're near a Wi-Fi hotspot, your PC should automatically pick up the wireless signal. Just make sure your computer's Wi-Fi adapter is turned on (it should be, by default), and then get ready to connect.

1 Click or tap the middle of the notification area of the taskbar to display the Quick Settings panel.

2 Click or tap the right arrow on the Manage Wi-Fi Connections button. You see a list of available wireless networks. (Public networks should not have lock icons next to their names.)

3 Click or tap to select the wireless network.

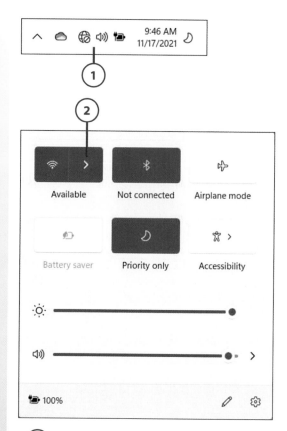

4 If the hotspot has free public access, click the Connect button. You can now open your web browser and surf normally.

5 If the hotspot requires a password, payment, or other logon procedure, Windows displays an Open Browser and Connect link. Click or tap this link to open your web browser and display the hotspot's logon page.

6 Enter the appropriate information or click the appropriate button to begin surfing.

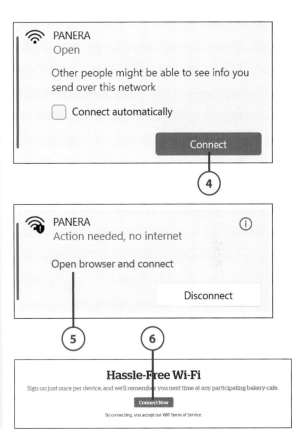

It's Not All Good

Unsecured Networks

Most public Wi-Fi hotspots are unsecured, which means that the information you send over these networks could be intercepted by others. For your security, you should avoid sending personal or financial information over unsecured public Wi-Fi hotspots. (To be safe, that means no online banking or shopping over public wireless networks.)

>>>Go Further

AIRPLANE MODE

If you're using your laptop or tablet on an airplane and don't want to use the plane's wireless Internet service (if available), you can switch to Airplane mode so that you can use your computer while in the air.

To switch into Airplane mode, click or tap the left side of the notification area of the taskbar to open the Quick Settings panel and then click to activate the Airplane Mode button. (It turns blue when activated.) You can switch off Airplane mode when your plane lands by clicking the blue Airplane Mode button.

If the plane you're on offers Wi-Fi service (many now do), you don't have to bother with Airplane mode. Instead, just connect to the plane's wireless network as you would to any Wi-Fi hotspot. You might have to pay for it, but it enables you to use the Internet while you're en route—which is a great way to spend long trips!

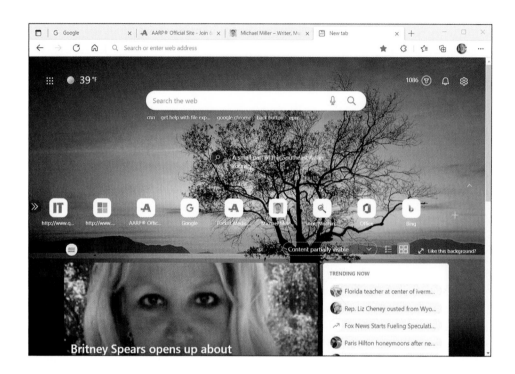

In this chapter, you find out how to use a web browser to browse and search pages on the World Wide Web.

→ Understanding the Web
→ Using Microsoft Edge
→ Searching the Internet

Browsing and Searching the Web

After you've connected to the Internet, either at home or via a public wireless hotspot, it's time to get busy. The World Wide Web is a particular part of the Internet with all sorts of cool content and useful services, and you surf the Web with a piece of software called a *web browser*.

Windows 11 includes a web browser, called *Microsoft Edge*, that you use to browse the Web—as well as search it for fun and useful information.

Understanding the Web

Before you can surf the Web, it helps to understand a little bit about how it works.

Information on the World Wide Web is presented in pages. A *web page* is similar to a page in a book, made up of text and graphics. A web page differs from a book page, however, in that it can include other elements, such as audio and video, as well as links to other web pages.

It's this linking to other web pages that makes the Web such a dynamic way to present information. A *link* on a web page can point to another web page on the same site or to another site. Most links are included as part of a web page's text and are technically called *hypertext links*, or just *hyperlinks*. (If a link is part of a graphic, it's called a *graphic link*.) These links are usually in a different color from the rest of the text and often are underlined; when you click or tap a link, you're taken directly to the linked page.

Web pages reside at a website. A *website* is nothing more than a collection of web pages (each in its own computer file) residing on a host computer. The host computer is connected full time to the Internet so that you can access the site—and its web pages—any time you access the Internet. The main page at a website is called the *home page*, and it often serves as an opening screen that provides a brief overview and menu of everything you can find at that site. The address of a web page is called a URL, which stands for *uniform resource locator*. Most URLs start with http:// (or https://, for secure sites), add www., continue with the name of the site, and end with .com, .org, or .net.

>>>*Go Further*
DIFFERENT TYPES OF ADDRESSES

Don't confuse a web page address with an email address or with a traditional street address. They're not the same.

As noted previously, a web page address has three parts, each separated by a period. A web page address typically starts with a www, followed by the domain name, and ends with a com, edu, gov, net, org, or other domain name extensions, and there are no spaces in it. You can use either lowercase or uppercase letters when you enter a web page address into the Address box in your web browser.

An email address also consists of three parts. The first part of the address, separated by an "at" sign (@), is your personal identifier, often your own name. After the @ sign is your email provider's domain, followed by a .com or similar domain extension. A typical email address looks something like this: *yourname@email.com*. You use email addresses to send emails to your friends and family; you do *not* enter email addresses into the Address box in your web browser.

Of course, you still have a street address that describes where you physically live. You don't use your street address for either web browsing or email online; it's solely for postal mail, and for putting on the front of your house or apartment.

No http://

You can normally leave off the http:// when you enter an address into your web browser. In most cases, you can even leave off the www. and just start with the domain part of the address.

Using Microsoft Edge

Microsoft includes the Microsoft Edge web browser in Windows 11. You can use Microsoft Edge for all your web browsing.

Open and Browse Web Pages

You launch Microsoft Edge from the Start menu. Once launched, you can use Edge to visit any page on the Web.

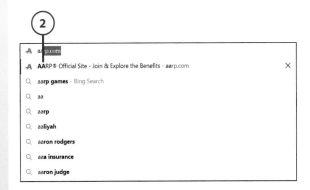

1. Start to type a web page address into the Address box.

2. As you type, Edge displays a list of suggested pages. Click or tap one of these pages or finish entering the web page address and press Enter.

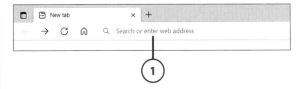

3. Click or tap the Back (left arrow) button beside the Address box to return to the previous web page.

4. Click or tap the Forward (right arrow) button to move forward again.

5. Click or tap the Refresh button to reload or refresh the current page.

6. Click or tap the home button to return to Edge's home page.

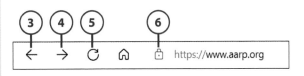

⑦ Pages on the Web are linked via hyperlinks, typically presented with colored or underlined text. Click or tap on a link to display the linked-to page.

Larger Text

If the text on a given web page is too small for you to read, Microsoft Edge lets you zoom in to (or out of) the page. Click or tap the Settings and More (three-dot) button (or press Alt+F on your computer keyboard), go to the Zoom section, and click + to enlarge the page (or click – to make the page smaller).

Work with Tabs

Most web browsers, including Microsoft Edge, let you display multiple web pages as separate tabs, and thus easily switch between web pages. This is useful when you want to reference different pages or want to run web-based applications in the background.

① Click or tap the + next to the last open tab to open a new tab. (Alternatively, press Ctrl+T on your computer keyboard.) A new home page opens.

② Enter a URL into the Address box, select from one of your previously visited pages, or start a search from the Search the Web box.

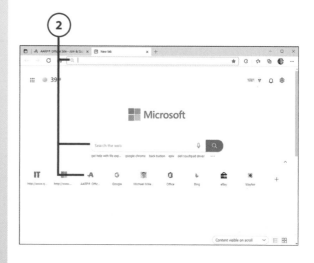

3 Switch tabs by clicking or tap-ping the tab you want to view. (Alternatively, press Ctrl+Tab to move to the next tab.)

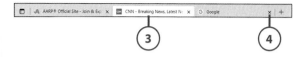

4 Click or tap the X on a tab to close it.

Save Favorite Pages

All web browsers let you save or bookmark your favorite web pages. In Microsoft Edge, you do this by adding pages to the Favorites list.

1 Navigate to the web page you want to add to your Favorites list and then click or tap the Favorites (star) icon in the Address box.

2 Confirm or enter a name for this page in the Name box.

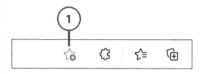

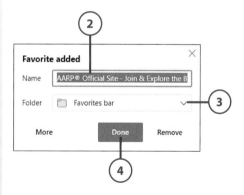

3 Favorites can be organized in folders. Pull down the Folder list to determine where you want to save this favorite.

4 Click or tap the Done button.

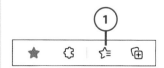

Return to a Favorite Page

To return to a page you've saved as a favorite, open the Favorites list and make a selection.

1 Click or tap the Favorites button to display the Favorites panel.

2 Click or tap the page or set of tabs you want to revisit.

Favorites Bar

For even faster access to your favorite pages, display the Favorites bar at the top of the browser window beneath the Address bar. Click the Settings and More button, click Favorites, click Show Favorites Bar, and then click Always.

Favorites

▲ ☆ Favorites bar

   A AARP® Official Site - Join & Explore the Benefits

   Y! My Yahoo!

   . TVparty! - Classic TV

   PD Weather Blog

   BringMeTheNews

   PP Pioneer Press — **2**

   ★ StarTribune

   13 WTHR

   wp Washington Post

Revisit History

Microsoft Edge makes it easy to see what pages you've recently visited—and return to any of those pages.

1 Click or tap the Settings and More (three-dot) button (or press Alt+F) to display the pull-down menu.

2 Click or tap History.

New tab		Ctrl+T
New window		Ctrl+N
New InPrivate window		Ctrl+Shift+N
Zoom	— 100% +	↗
Favorites		Ctrl+Shift+O
Collections		Ctrl+Shift+Y
History		Ctrl+H
Downloads		Ctrl+J

(3) Pages are displayed in reverse chronological order. Click or tap a page to reopen it.

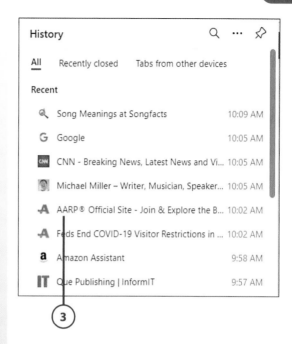

Browse in Private

If you want to browse anonymously, without any traces of your history recorded, activate InPrivate Browsing mode in a new browser window. With InPrivate Browsing, no history is kept for the pages you visit, so no one can track where you've been.

(1) Click or tap the Settings and More (three-dot) button.

(2) Select New InPrivate Window.

(3) A new InPrivate Edge window opens, ready to accept any URL you input.

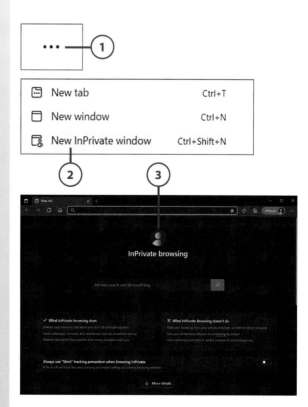

View a Page in the Immersive Reader

Some web pages are overly cluttered with advertisements and other distracting elements. You can get rid of these visual distractions by using Edge's Immersive Reader, which also increases the size of the text on the page.

With the Immersive Reader, all the unnecessary items on a page are removed, so all you see is the main text and accompanying pictures. In addition, the Immersive Reader makes all onscreen text significantly larger, and there's more "white space" all around. The result is, perhaps, the best way to view web pages if you have even slight vision difficulties.

Not All Pages

Not all web pages are compatible with the Immersive Reader. If the page can't be viewed with the Immersive Reader, the Immersive Reader icon is not displayed.

① Navigate to the page you want to read; then click or tap the Enter Immersive Reader icon in the Address box. (Alternatively, you can enter immersive reading mode by pressing F9 on your keyboard.)

② The page displays without the unnecessary elements and increased text size for easier reading. Scroll through the page as normal to read the entire article.

③ The Immersive Reader can also use text-to-speech technology to read the page to you. Move your mouse to the top of the page to display the menu of options; then click or tap Read Aloud.

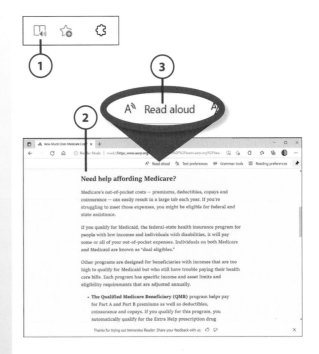

4 Click or tap Pause to pause the reading. Click or tap Play to resume.

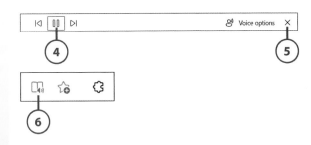

5 Click or tap the X to exit Read Aloud mode.

6 Click or tap the Exit Immersive Reader icon in the Address bar (or press F9 again) to return to normal view.

Print a Web Page

From time to time, you might run across a web page with important information you want to keep for posterity. While you can make this page a favorite, of course, you can also print it using your computer's printer.

1 Click or tap the Settings and More (three-dot) button.

2 Click or tap Print to display the Print window.

3 Click or tap the Printer list and select your printer.

4 Enter how many copies you want to print.

5 Click or tap to select either Portrait or Landscape printing mode.

6 Click or tap the Pages list and select which pages you want to print (all or a range of pages).

7 Click or tap the Print button to print a copy of this page.

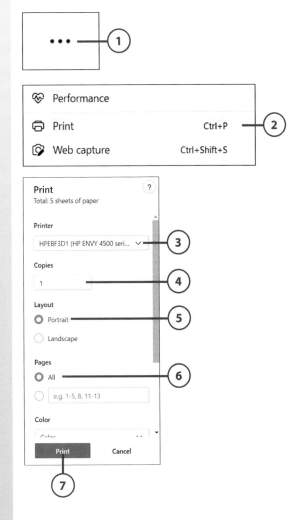

Set Your Home Page

Microsoft Edge lets you set a home page that can automatically open whenever you launch the browser or when you click or tap the browser's Home button. By default, Edge displays the New Tabs page, which you can customize with different layouts. You can also select a specific page on the Web to open as your home page. (For example, you might want to set a news site, such as CNN, as your home page.)

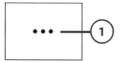

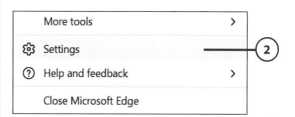

1. Click or tap the Settings and More (three-dot) button to open the pull-down menu.

2. Click or tap Settings to open the Settings page.

3. In the left column, click or tap Start, Home, and New Tabs.

4. In the Home Button section, select either New Tab page or Enter URL to open a different page when you click or tap the home button.

5. If you select Enter URL, enter the URL of the page you want as your home page; then click or tap Save.

Select Which Pages Open When You Launch Edge

You can also set which page or pages (on separate tabs) open when you launch the Edge browser. By default, Edge opens the New Tab page, but there are other options.

1. Click or tap the Settings and More (three-dot) button to open the pull-down menu.

2. Click or tap Settings to open the Settings page.

3. In the left column, click or tap Start, Home, and New Tabs.

4. In the When Edge Starts section, select one of the following: Open the New Tab Page, Open Tabs from the Previous Session, or Open These Pages.

5. If you selected Open These Pages, click or tap Add a New Page and then select which page(s) you want to open when you launch Edge.

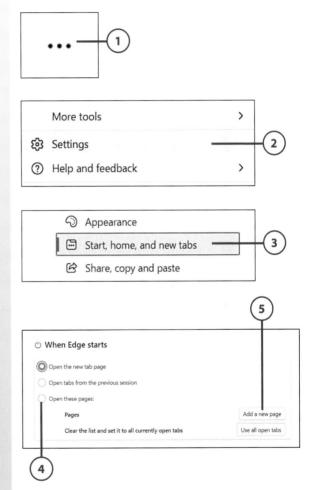

Configure the New Tab Page

By default, the New Tab page is a rather bland and empty page that displays a search box and icons for the last few web pages you've visited. You can make the page look more appealing if you want, as well as display other information.

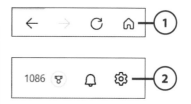

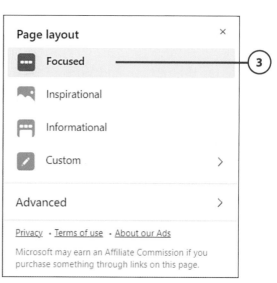

1 Click or tap the home button to open the current home page. By default, this is the New Tab page.

2 Click or tap the Settings (gear) button on the New Tab page.

3 Select a different layout for the New Tabs page—Focused, Inspirational, Informational, or Custom.

4 If you selected Custom, you have additional choices. You can select how many rows of "quick links" (most recent web pages visited) to display; whether to display Bing searches, a greeting, or a games sidebar; and what background you'd like to see. You can also opt to show additional content in several ways—(all) content visible, content partially visible, headings only, or content off.

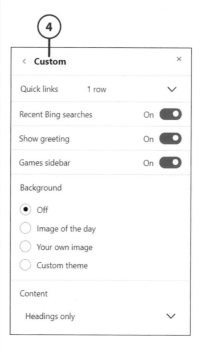

Page layout ✕

▪▪▪ Focused ──────── 3

💬 Inspirational

🛏 Informational

✏ Custom >

Advanced >

Privacy · Terms of use · About our Ads

Microsoft may earn an Affiliate Commission if you purchase something through links on this page.

4

< **Custom** ✕

Quick links   1 row   ∨

Recent Bing searches   On ⬤

Show greeting   On ⬤

Games sidebar   On ⬤

Background

● Off

○ Image of the day

○ Your own image

○ Custom theme

Content

Headings only   ∨

>>>*Go Further*

USING GOOGLE CHROME AND OTHER WEB BROWSERS

Microsoft Edge isn't the only web browser you can use to surf the Internet. Several other browsers are available.

The most popular web browser today is Google Chrome, which you can download for free here: www.google.com/chrome/. Chrome is built on the same engine as the new Microsoft Edge, features similar functionality, and has a similar interface.

Also popular are the Apple Safari (www.apple.com/safari) and Mozilla Firefox (www.mozilla.org/firefox) browsers. Safari is the default browser on Apple's Mac computers and iPhone and iPad devices but isn't supported on Windows computers. Firefox has a core base of dedicated users but hasn't achieved widespread usage.

Searching the Internet

There is so much information on the Web—so many web pages—that it's sometimes difficult to find exactly what you're looking for. The best way to find just about anything on the Internet is to search for it, using a *web search engine*.

Search Google

The most popular search engine today is Google (www.google.com), which indexes billions of individual web pages. Google is easy to use and returns extremely accurate results.

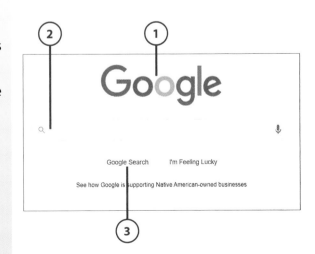

1. From within your web browser, type **www.google.com** into the Address box and then press Enter. This opens Google's main search page.

2. Enter one or more keywords into the Search box.

3. Press Enter or click or tap the Google Search button.

④ When the results are displayed, go to the result you want to view and then click or tap the link for that result. This displays the selected web page within your web browser.

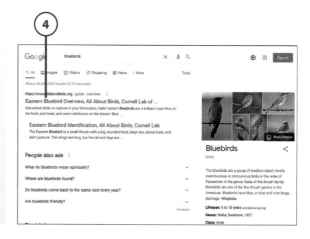

>>>Go Further

FINE-TUNE YOUR SEARCH RESULTS

Google lets you fine-tune your search results to display only certain types of results. Use the links at the top of Google's search results page to display only Images, Videos, Maps, Shopping, or News results—or click or tap Search Tools for further refinements.

Search Bing

Microsoft has its own search engine, called Bing (www.bing.com). It works pretty much like Google, and Microsoft would very much like you to use it.

① From within your web browser, enter www.bing.com into the Address box and press Enter. This opens Bing's main search page.

② Enter one or more keywords into the Search box.

③ Press Enter or click or tap the Search (magnifying glass) button.

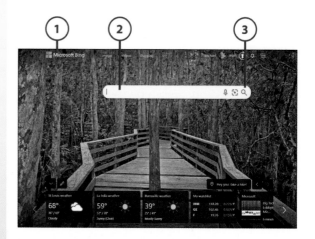

4 When the results are displayed, click any page link to view that page.

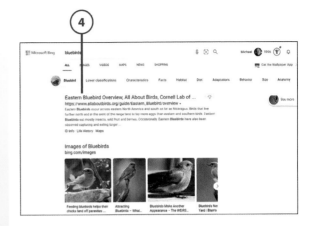

Search DuckDuckGo

Another search engine gaining in popularity is called DuckDuckGo. This search engine's claim to fame is that it doesn't track your browsing or retain your search history, as both Google and Bing do.

1 From within your web browser, enter www.duckduckgo.com into the Address box and press Enter. This opens DuckDuckGo's main search page.

2 Enter one or more keywords into the Search box.

3 Press Enter or click or tap the Search (magnifying glass) button.

4 When the results are displayed, click or tap any page link to view that page.

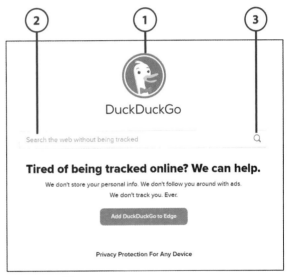

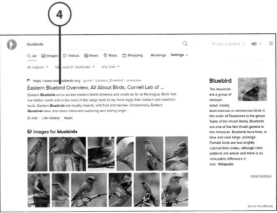

Change the Default Search Engine

Microsoft Edge, like Google Chrome and other web browsers, lets you initiate searches directly from its Address box. When you enter one or more words into the Address box, Edge sends your search query to the Bing search engine. (If you do this in Google Chrome, Chrome sends your query to the Google search engine.)

If you'd rather use Google or DuckDuckGo instead of Bing, you can easily change Edge's built-in search function to default to the other search engine.

1. From within Microsoft Edge, click or tap the Settings and More (three-dot) button.

2. Click or tap Settings to open the Settings window.

3. In the left column, click or tap Privacy, Search, and Services.

4. Scroll to the very bottom of the page and click or tap Address Bar and Search.

5. Click or tap the Search Engine in the Address Bar list and make another selection.

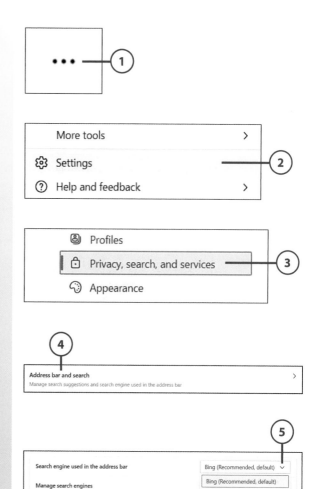

In this chapter, you learn how to safely shop online—including ordering meals and groceries for delivery.

→ Purchasing from Online Retailers

→ Buying and Selling at Online Marketplaces

→ Ordering Meals, Groceries, and More for Delivery

12

Shopping and Ordering Online

Online shopping has long been a practical alternative for people who find real-world shopping inconvenient, at best. During the COVID-19 crisis, more and more people turned to online shopping and online ordering for meals, groceries, and more. Almost all businesses are online these days, which makes shopping and ordering from your computer an easy thing to do.

Purchasing from Online Retailers

If you've never shopped online before, you're probably wondering just what to expect. Shopping online is actually easy; all you need is your computer and a credit card—and a fast connection to the Internet!

The online shopping experience is similar from retailer to retailer. You typically proceed through a multiple-step process from discovery to ordering to checkout and payment.

Discover Online Retailers

The first step in online shopping is finding where you want to shop. Most major retailers, such as Best Buy (www.bestbuy.com), Costco (www.costco.com), The Home Depot (www.homedepot.com), Kohl's (www.kohls.com), Macy's (www.macys.com), Office Depot (www.officedepot.com), Nordstrom (www.nordstrom.com), Sephora (www.sephora.com), Target (www.target.com), and Walmart (www.walmart.com), have websites you can use to shop online. Most catalog merchants, such as Coldwater Creek (www.coldwatercreek.com), L.L. Bean (www.llbean.com), and Land's End (www.landsend.com), also have websites for online ordering.

In addition, there are many online-only retailers that offer a variety of merchandise. These are companies without physical stores; they conduct all their business online and then ship merchandise direct to buyers. These range from smaller niche retailers to larger full-service sites, such as Amazon.com (www.amazon.com), Overstock.com (www.overstock.com), and Wayfair (www.wayfair.com).

Search or Browse for Merchandise

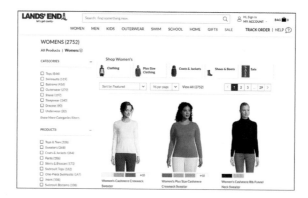

After you've determined where to shop, you need to browse through different product categories on that site or use the site's search feature to find a specific product.

Browsing product categories online is similar to browsing through the departments of a retail store. You typically click a link to access a major product category, and then click further links to view subcategories within the main category. For example, the main category might be Clothing; the subcategories might be Men's, Women's, and Children's clothing. If you click the Men's link, you might see a list of further subcategories: Outerwear, Shirts, Pants, and the like. Just keep clicking until you reach the type of item that you're looking for.

When you have something specific in mind, searching for products is often a faster way to find what you're looking for. For example, if you're looking for a women's sweater, you can enter the words **women's sweater** into the site's search box and get a list of specific items that match those criteria. You can often get even more specific in your search by using a filter the merchant offers to specify a particular color, size, brand, and more.

The only problem with searching is that you might not know exactly what it is you're looking for; if this describes your situation, you're probably better off browsing. But if you *do* know what you want—and you don't want to deal with lots of irrelevant items—then searching is the faster option.

Examine the Product (Virtually)

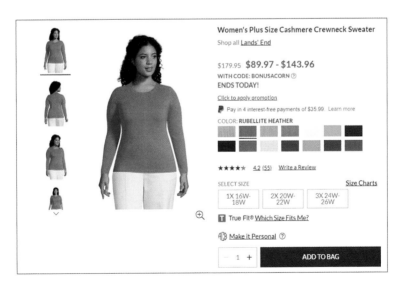

Whether you browse or search, you'll probably end up looking at a list of different products on a web page. These listings typically feature one-line descriptions of each item—in most cases, not nearly enough information for you to make an informed purchase.

The thing to do now is to click the link for the item you're particularly interested in. This should display a dedicated product page, complete with a picture and full description of the item. This is where you can read more about the item you selected. Some product pages include different views of the item, pictures of the item in different colors or sizes, links to additional information, customer reviews, and maybe even a list of optional accessories that go along with the item.

Many retailers feature customer ratings and reviews of their products. Read these reviews to see what other customers liked or disliked about a given product—including, with clothing items, whether they run big, small, or true to size.

If you like what you see, you can proceed to the ordering stage. If you want to look at other items, just click your browser's Back button to return to the larger product listing.

Make a Purchase

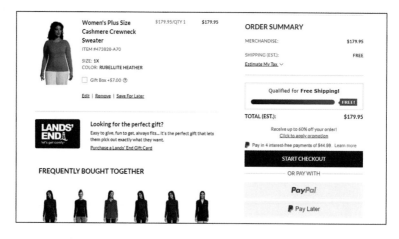

Somewhere on each product description page should be a button labeled Purchase, Buy Now, Add to Cart, Add to Bag, or something similar. This is how you make the purchase: by clicking that "buy" button. You don't order the product just by looking at the product description; you have to manually click the "buy" button to place your order.

When you click the "buy" button, that particular item is added to your *shopping cart*. That's right, the online retailer provides you with a virtual shopping cart that functions just like a real-world shopping cart. Each item you choose to purchase is added to your virtual shopping cart.

After you've ordered a product and placed it in your shopping cart, you can choose to shop for other products on that site or proceed to the site's *checkout*. It's important to note that when you place an item in your shopping cart, you haven't actually completed the purchase yet. You can keep shopping (and adding more items to your shopping cart) as long as you want.

You can even decide to abandon your shopping cart and not purchase anything at this time. All you have to do is leave the website, and you won't be charged for anything. It's the equivalent of leaving your shopping cart at a real-world retailer and walking out the front door; you don't actually buy anything until you walk through the checkout line. (Although, with some sites, the items remain in your shopping cart—so they'll be there waiting for you the next time you shop!)

Check Out and Pay

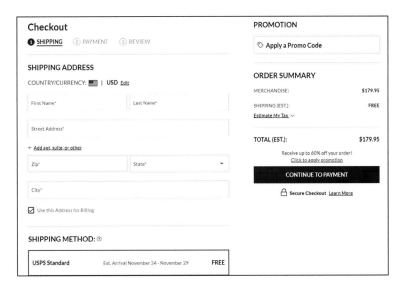

To finalize your purchase, you have to visit the store's checkout. This is like the checkout line at a traditional retail store; you take your virtual shopping cart through the checkout, get your purchases totaled, and then pay for what you're buying.

The checkout at an online retailer typically consists of one or more web pages with forms you have to fill out. If you've visited the retailer before, the site might remember some of your personal information from your previous visit. Otherwise, you have to enter your name, address, and phone number, as well as the address you want to ship the merchandise to (if it's different from your billing address—for example, if you are sending a gift to someone). You also have to pay for the merchandise, typically by entering a credit card number.

The checkout provides one last opportunity for you to change your order. You can delete items you decide not to buy or change quantities on any item. At some merchants, you can even opt to have your items gift-wrapped and sent to someone as a present. You should be able to find all these options somewhere in the checkout process.

You might also have the option of selecting different types of shipping for your order. Many merchants offer both regular and expedited shipping—the latter for an additional charge. It's also common to find free shipping on orders above a certain dollar amount; sometimes it makes sense to spend a few dollars more to get that free shipping.

Another option at some retailers is to group all items for reduced shipping cost. (The alternative is to ship items individually as they become available.) Grouping items is attractive cost-wise, but you can get burned if one of the items is out of stock or not yet available; you could end up waiting weeks or months for those items that could have been shipped immediately.

After you've entered all the appropriate information, you're asked to place your order. This typically means clicking a button that says "Place Your Order" or something similar. You might even see a second screen asking you whether you *really* want to place your order, just in case you have second thoughts.

After you place your order, you see a confirmation screen, typically displaying your order number. Write down this number or print this page; you need to refer to this number if you have to contact customer service. Most online merchants also send you a confirmation message, including this same information, via email.

That's all there is to it. You shop, examine the product, place an order, proceed to checkout, and pay—then your order should arrive within the designated timeframe. It's that easy!

It's Not All Good

How to Shop Safely Online

Some consumers are wary about buying items online, even though shopping online can be every bit as safe as shopping at your local retail store. The big online retailers are just as reputable as traditional retailers, offering safe payment, fast shipping, and responsive service.

To make online shopping as safe as possible, take the following precautions:

- Make sure the online retailer prominently displays its contact information and offers multiple ways to connect. You want to be able to email, call, tweet, or text chat with the retailer if something goes wrong.

- Look for the site's return policy and satisfaction guarantee. You want to be assured that you'll be taken care of if you don't like what you ordered.

- A reputable site should tell you whether an item is in stock and how long it will take to ship—before you place your order. Many stores let you order online and then pick up at a nearby location.

- Purchase only from retailers that use a secure server for the checkout process. Look in the Address box for https:// (not the normal http://) before the URL; you should also see the "lock" symbol before or after the address. If the checkout process is not secure, do not proceed with payment.

- For the best protection, pay by major credit card. (You can always dispute your charges with the credit card company if something goes wrong and the retailer won't make it good.)

Indeed, the safest way to shop online is to pay via credit card because credit card purchases are protected by federal law. In essence, you have the right to dispute certain charges, and your liability for unauthorized transactions is limited to $50. In addition, some card issuers offer a supplemental guarantee that says you're not responsible for *any* unauthorized charges made online. (Read your card's statement of terms to determine the company's exact liability policy.)

Buying and Selling at Online Marketplaces

Traditional retailers aren't the only places to buy merchandise online. You can also buy goods from individuals selling to other individuals through a number of online marketplaces. These marketplaces also let you sell things yourself, which is a good way to move things you no longer use, much like an online garage sale.

Craigslist

Craigslist (www.craigslist.org) works like an online version of traditional newspaper classified ads. Individuals list items for sale, and you contact those sellers (via Craigslist) to arrange purchases.

Craigslist only does ad listings, so you have to arrange payment directly to the seller. (Most sellers accept cash only.) Unlike some online marketplaces, Craigslist is best for buying and selling locally, where you can pick up the items you buy directly from the sellers; it's not really for buying or selling items that need to be shipped.

Craigslist also lets you place listings for items you have for sale. In most categories, listings are free.

It's Not All Good

Buyer Beware

Just as with traditional classified ads, Craigslist offers no buyer protections. Before handing over any money, plug in anything electric or electronic and test its capabilities, thoroughly inspect items in good lighting and from all angles, and make sure the product is exactly what you want. You should also arrange to pick up any items you buy in a public space and take a friend with you, for extra protection.

eBay

eBay (www.ebay.com) started out as an online auction site but today offers a mixture of items for auction or for sale at a fixed price. In an online auction, you bid for a given item, and the buyer with the highest bid at the end of the auction period wins the item. Fixed-price sales are just like buying an item from a normal online retailer.

When you purchase from an eBay seller, you make your payments through eBay and are covered by eBay's buyer protection plan. Note, however, that when you're buying from individual sellers, it's not always as smooth or as safe as buying from a normal retailer. (Many individuals selling on eBay don't accept returns, for example.)

Unlike Craigslist, which is mainly for local sales, eBay sellers sell and ship items across the United States and around the world. In addition to individual sellers, many traditional retailers and online retailers offer products for sale on the eBay marketplace.

You can also sell your own items on eBay, either for auction or for a fixed-price. You pay eBay a fee to list the item and another fee when the item sells.

Etsy

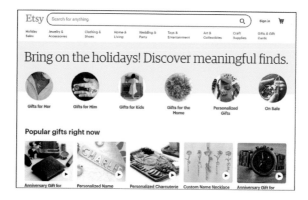

Etsy (www.etsy.com) is a marketplace for handmade and vintage items such as artwork, clothing, collectibles, crafts, jewelry, and such. Sellers are often individuals who make their own goods and sell them online via Etsy. You pay via the Etsy site using credit card, debit card, or PayPal.

Etsy is a great site to sell items that you make yourself. Like eBay, you pay a listing fee to list an item for sale and a transaction fee when the item sells.

Facebook Marketplace

If you're a Facebook member, you can buy and sell items on the Facebook Marketplace (www.facebook.com/marketplace). The Facebook Marketplace is a lot like Craigslist in that it's really just a series of item listings. You don't buy and sell through Facebook; you just use Facebook to list items for sale.

If you want to buy an item, you contact the seller (another Facebook member) directly and arrange payment and pickup with them. (Cash is king on the Facebook Marketplace.) As with Craigslist, buying something listed on the Facebook Marketplace is strictly buyer beware.

Selling on the Facebook Marketplace means creating a listing for the item you want to sell. There are no fees involved.

Facebook

Facebook itself is a social network you can use to connect with family and friends. Learn more about Facebook in Chapter 17, "Connecting with Facebook and Other Social Media."

Reverb

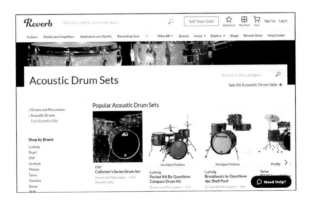

Reverb (www.reverb.com) is an online marketplace for new, used, and vintage musical instruments, DJ gear, and recording and sound reinforcement equipment. It's the go-to site for musicians and anyone working with them.

You can find gear from both individual sellers and music stores on Reverb. You pay through Reverb via credit/debit card or PayPal. The site offers a buyer protection plan, although buying from individual sellers can sometimes be problematic if you end up with a defective item or something is not as described.

Reverb is a pretty good place for selling used musical instruments and gear because you have a targeted audience. While Reverb doesn't charge listing fees, you do pay a 5% fee when you sell an item, along with a 2.5% to 2.7% payment processing fee.

Ordering Meals, Groceries, and More for Delivery

During that period of the COVID-19 crisis when we were all stuck at home, people used their computers not just for online shopping but for ordering meals, groceries, and other items to be delivered. Many people got really used to convenience of ordering their food and other sundries online and having them delivered direct to their doors. You never have to leave home again.

Most grocery stores, pharmacies, and restaurants either offer their own delivery services or partner with local or national delivery services. You typically order from the grocery or restaurant website and then choose the delivery option.

Order Meals Online

Many restaurants today offer delivery. This delivery is seldom free; you'll have to pay a delivery fee and perhaps even specify a tip to the driver. All of these fees— as well as the cost of the meal—are paid online when you place your order. You can typically pay via credit or debit card or via PayPal.

Some restaurants hire their own drivers and do their own deliveries. Others employ the services of third-party services such as DoorDash (www.doordash.com), Grubhub (www.grubhub.com), and UberEats (www.ubereats.com).

You can order direct from the restaurant or, in some cases, from the delivery site. After you place your order and pay online, the site will tell you approximately when your order will be delivered. You may also receive an email or see a link to an online page that displays the status of your order and sometimes even a map with the position of the delivery driver highlighted.

Many restaurants let you choose normal or contactless delivery. With a normal delivery, the driver rings the doorbell and hands you your food. With contactless delivery, the driver places the food on your doorstep and texts you that it's there, so there's no human contact involved. You make this choice when you're placing your order.

Order Groceries Online

Many grocery stores also offer online ordering and home delivery. The process works much like ordering meal delivery from a restaurant but with many more options.

In most instances, you place your order directly from the grocer's website. Most grocery stores offer the same selection online as they do in their stores, including fresh meats, vegetables, bakery goods, and dry goods. You may be prompted to specify options in case your first choice of brand or size isn't available, but it's pretty much like ordering from a big menu of available grocery items.

Many larger grocery chains handle their own deliveries, but others use third-party delivery services such as InstaCart (www.instacart.com) or Shipt (www.shipt.com). You'll probably have to pay a delivery fee and perhaps specify a tip for the delivery driver. Because many grocery items need to stay refrigerated, most stores let you specify a delivery time so you can be sure that you're home to receive the delivery.

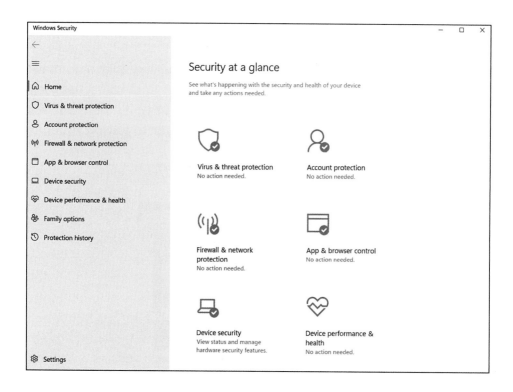

In this chapter, you become familiar with the most common online threats and find out how to protect yourself against them.

→ Protecting Against Identity Theft and Phishing Schemes
→ Protecting Against Online Fraud
→ Protecting Against Computer Viruses and Other Malware

13

Protecting Yourself Online

While most sites on the Internet are safe, there are some unscrupulous operators waiting to prey on unsuspecting users. You can, however, take steps to protect yourself when you're online. You need to be able to identify the most common online threats and scams and know how to avoid becoming a victim.

Protecting Against Identity Theft and Phishing Schemes

Online predators want your personal information—your real name, address, online usernames and passwords, bank account numbers, Social Security numbers, and the like. It's called *identity theft*, and it's a way for a con artist to impersonate you—both online and in the real world. If your personal data falls into the hands of identity thieves, it can be used to hack into your online accounts, make unauthorized charges on your credit card, drain your bank account, and more.

Identity theft is a major issue. Close to 1.4 million cases of identity theft were reported to the FTC in 2020, an increase of more than 50% over the prior year. A third of those identity theft victims were male, and two-thirds were female. The two most heavily targeted age groups were people aged 30 to 39 and 40 to 49, with victims aged 50 to 59 ranking third.

A case of identity theft costs the average victim more than $1,000. As a result of the crime, victims reported difficulties paying their rent or mortgage, paying utilities, and buying necessities like groceries.

Criminals have many ways to obtain your personal information. Almost all involve tricking you, in some way or another, into providing this information of your own free will. Your challenge is to familiarize yourself with their tricks so you can avoid becoming a victim.

Avoiding Phishing Scams

Online, identity thieves often use a technique called *phishing* to trick you into disclosing valuable personal information. It's called that because the other party is "fishing" for your personal information, typically via fake email messages and websites.

It's Not All Good

Phishing **Means** *Phony*

A phishing scam typically starts with a phony email or text message that appears to be from a legitimate source, such as your bank, the postal service, PayPal, or other official institution. This message purports to contain important information that you can see if you click the enclosed link. That's where the bad stuff starts.

If you click the link in the phishing message, you're taken to a fake website masquerading as the real site, complete with logos and official-looking text. You're encouraged to enter your personal information into the forms on this fake web page; when you do so, your information is sent to the scammer, and you're now a victim of identity theft.

How can you avoid falling victim to a phishing scam? There are several things you can do:

- Look at the sender's email address. Most phishing emails come from an address different from the one indicated by the (fake) sender. For example, in an email that's supposedly from FedEx, the email address 619.RFX@jacksonville.com would be suspicious; you'd expect an email from FedEx to look something like *address*@fedex.com. Even trickier are those addresses that closely resemble an official address, such as the fake *address*@fedexmail.com instead of the real *address*@fedex.com.

- Mouse over any links in the email. In a phishing email, the URL for the link will not match the link text or the (fake) sender's supposed website.

- Look for poor grammar and misspellings. Many phishing schemes come from outside the United States by scammers who don't speak English as their first language. As such, you're likely to find questionable phrasing and unprofessional text—not what you'd expect from your bank or other professional institution.

- If you receive an unexpected email, no matter the apparent source, do *not* click any of the links in the email. If you think there's a legitimate issue from a given website, go to that site manually in your web browser and access your account from there.

- Some phishing messages include attached files that you are urged to click to display a document or image. Do *not* click or open any of these attachments; they might contain malware that can steal personal information or damage your computer. (Read more about malware later in this chapter.)

- Not all phishing scams come via email. You should also beware of text messages from people you don't know, as well as scam direct messages on Facebook, Twitter, and other social media.

Phishing Filters

Many web browsers—including both Google Chrome and Microsoft Edge—offer some built-in protection against phishing scams in the form of filters that alert you to potential phishing sites. If you click a bad link or attempt to visit a known or suspected phishing site, the browser displays a warning message. Do not enter information into these suspected phishing sites—return to the previous page instead!

Keeping Your Private Information Private

Identity theft can happen any time you make private information public. This has become a special issue on social networks, such as Facebook, where users tend to forget that everything they post is publicly visible.

Many Facebook users not only post personal information in their status updates, but also include sensitive data in their personal profiles. Javelin Strategy and Research found that 68% of people with public social media profiles shared their birthday information, 63% shared the name of their high schools, 18% shared their phone numbers, and 12% shared their pet's names.

None of this might sound dangerous, until you realize that all of these items are the type of personal information many companies use for the "secret questions" their websites use to reset users' passwords. A fraudster armed with this publicly visible information could log on to your account on a banking website, for example, reset your password (to a new one he provides), and thus gain access to your banking accounts.

The solution to this problem is to enter as little personal information as possible when you're online. For example, you don't need to—and shouldn't—include your street address or phone number in a comment or reply to an online news article. Don't give the bad guys anything they can use against you!

Follow these tips:

- Unless absolutely necessary, do not enter your personal contact information (home address, phone number, and so on) into your social media profile.

- Do not post or enter your birthdate, children's names, pet's names, and the like—anything that could be used to reset your passwords at various websites.

- Do not post status updates that indicate your current location—especially if you're away from home. That's grist for both physical stalkers and home burglars.

- Don't play those "quizzes" on Facebook that ask seemingly innocuous questions (such as, "What was the name of your first grade teacher?") to complete a personality profile or something similar. These are often some of the questions you're asked if you need to reset your password on various sites; armed with this information, scammers can reset your password and take over your account without you knowing.

It's Not All Good

Corporate Data Breaches

As careful as you may be in protecting your personal information, the fact remains that many big companies collect information about you—and that information is not always secure. Many big-time hackers target large corporations—retailers, banks, financial firms—in search of data to steal, which can then be used to perpetrate identity theft and other crimes.

What can do you when your personal data is accessed by cybercriminals? Most companies that have been hacked offer ways to check on your personal data, including offering free credit monitoring for a period of time. You can also consider putting a credit freeze on your accounts so that unauthorized persons cannot open new accounts in your name. You should also manually monitor the transactions in your bank and credit card accounts so that you're immediately aware if something is amiss.

Unfortunately, there's not much you can do to prevent this sort of situation from happening; that responsibility lies in the hands of those storing your data. Unless you go completely off the grid, your personal data is going to be available to various companies online, and anything online (or even offline) can be hacked. We can hope that the big data companies get better at protecting our data, but until then, we just have to be alert to what happens if or when these criminal attacks occur.

Hiding Personal Information on Facebook

Too many Facebook users of all ages make all their personal information totally public—visible to all users, friends or not. Fortunately, you can configure Facebook's privacy settings to keep your private information private.

Table 13.1 details Facebook's available privacy settings.

Table 13.1   Facebook Privacy Settings

Setting	Who Can View
Public	Anyone on or off Facebook
Friends	People on your Facebook friends list
Friends except	People on your Facebook friends list *except* those you specify
Specific friends	Only those Facebook friends you specify
Only me	Nobody but you
Custom	Those friends and lists you specify

Facebook

Learn more about using the Facebook social network in Chapter 17, "Connecting with Facebook and Other Social Media."

(1) Click or tap the Account down arrow on the Facebook toolbar.

(2) Click or tap Settings & Privacy.

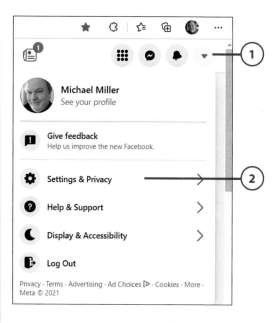

3 Click or tap Privacy Checkup.

4 Click or tap to select Who Can See What You Share.

5 Click or tap Continue to display the Profile Information screen.

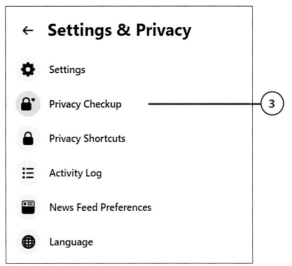

6 Scroll to a specific piece of information to view its privacy setting.

7 Click a privacy setting for a given item to change it.

8 Select who can see this item.

9 When you're returned to the Profile Information window, click Next to display the Posts and Stories page.

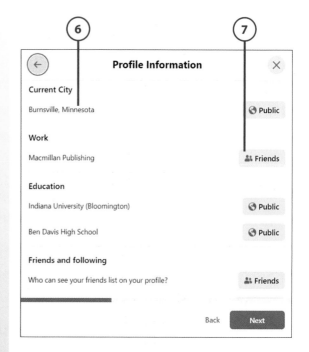

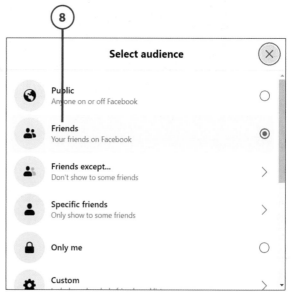

(10) Click to change the privacy settings for your future posts.

(11) Click to change the privacy settings for your stories.

(12) Click the Limit button to change who can view older posts.

(13) Click Next to display the Blocking screen.

(14) Click Add to Blocked List to block specific users from viewing your posts and profile information.

(15) Click Next.

(16) Click the X to close the window.

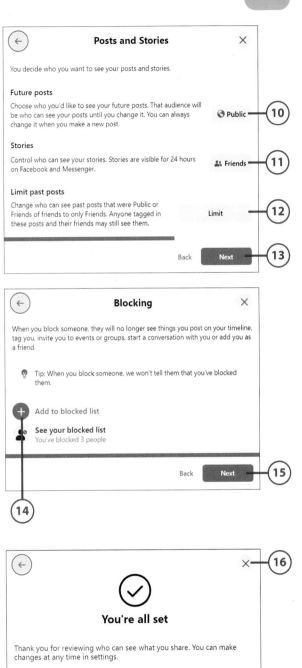

Protecting Against Online Fraud

Identity theft isn't the only kind of online fraud you might encounter. Con artists are especially creative in concocting schemes that can defraud unsuspecting victims of thousands of dollars.

Most of these scams start with an email message that promises something for nothing. Maybe the message tells you that you've won a lottery, or you are asked to help someone in a foreign country deposit funds in a U.S. bank account. You might even receive requests from people purporting to be far-off relatives who need some cash to bail them out of some sort of trouble.

The common factor in these scams is that you're eventually asked to either send money (typically via wire transfer) or provide your bank account information—with which the scammers can drain your money faster than you can imagine. The damage can be considerable.

Protecting yourself from the huge number of these online scams is both difficult and simple. The difficulty comes from the sheer number of scams and their amazing variety. The simplicity comes from the fact that the best way to deal with any such scam is to spot it and then ignore it.

Scams Are Not Spam

You can't rely on your email program's spam filter to stop scam emails. Spam and scams are two different things, even if they're both unwanted. Although some scam messages are stopped by spam filters, many messages get through the filter and land in your inbox, just as if they were legitimate messages—which, of course, they aren't.

>>>Go Further
AARP FRAUD WATCH NETWORK

Join AARP's Fraud Watch Network (www.aarp.org/fraudwatchnetwork) for free to learn how to protect yourself and your family from the latest scams and to report scams.

Identifying Online Scams

Most online fraud is easily detectable by the simple fact that it arrives in your email inbox out of the blue and seems too good to be true. So if you get an unsolicited offer that promises great riches, you know to hit the Delete key—pronto.

You can train yourself to recognize scam emails at a glance. That's because most scam messages have one or more of the following characteristics in common:

- The email does not address you personally by name; your name doesn't appear anywhere in the body of the message.

- You don't know the person who sent you the message; the message was totally unsolicited.

- The message is rife with spelling and grammatical errors. (Scammers often operate from foreign countries and do not speak English as their first language.) Conversely, the text of the message might seem overly formal, as if written by someone not familiar with everyday English.

- You are promised large sums of money for little or no effort on your part.

- You are asked to provide your bank account number, credit card number, or other personal information—or are asked to provide money upfront for various fees or to pay the cost of expediting the process.

- You are asked to buy one or more gift cards for your boss, pastor, or other authority figure—and then provide the cards' numbers or mail the cards to a strange address.

It's Not All Good

Tech Support Scams

Here's another scam that's going the rounds. You get a phone call from someone purporting to be from "Windows" or "Windows Technical Support." This person tells you he's received notice that your computer or version of Windows has been infected with viruses and that he can walk you through the steps to correct the problem.

The problem is that there isn't any problem; the person calling you has no knowledge of or connection to your computer, and if you go along with it, you're going to be scammed. The person on the other end of the phone may have you open a Windows tool called the Event Viewer to "show" you that you're under attack; you'll then see all the typical behind-the-scenes Windows alerts and warnings that look scary but are really quite normal. Once the person has convinced you that you have a problem, you're then conned into downloading and possibly paying for purported antimalware or tech support software that, in reality, places spyware and viruses on your computer. This malware might even require you to pay more and more, over time, for additional "fixes."

If you receive this type of phone call, just hang up.

Avoiding Online Fraud

Recognizing a scam email is just one way to reduce your risk of getting conned online. Here are some more tips you can employ:

- Familiarize yourself with the common types of online scams—and if a message in your inbox resembles any of these common scams, delete it.

- Ignore all unsolicited emails and text messages, of any type. No stranger will send you a legitimate offer via email or text; it just doesn't happen. When you receive an unsolicited offer via email, delete it.

- Don't give in to greed. If an offer sounds too good to be true, it probably is; there are no true "get rich quick" schemes.

- Never provide any personal information—including credit card numbers, your Social Security number, and the like—via email. If such information is legitimately needed, you can call the company yourself or visit its official website to provide the information directly.

>>>Go Further
WHAT TO DO IF YOU'VE BEEN SCAMMED

What should you do if you think you've been the victim of an email fraud? There are a few steps you can take to minimize the damage:

- If the fraud involved transmittal of your credit card or banking information, contact your credit card company or bank to halt all unauthorized payments—and to limit your liability.

- If you think your bank accounts have been compromised, contact your bank to put a freeze on your checking and savings accounts—and open new accounts, if necessary.

- Contact one of the three major credit-reporting bureaus to see if stolen personal information has been used to open new credit accounts—or max out your existing accounts. The three major bureaus are Equifax (www.equifax.com), Experian (www.experian.com), and TransUnion (www.transunion.com).

- Contact your local law enforcement authorities—fraud is illegal, and it should be reported as a crime.

- Report the fraud to your state attorney general's office.

- File a complaint with the Federal Trade Commission (FTC) via the form located at www.ftc-complaintassistant.gov.

- Contact any or all of the following consumer-oriented websites: Better Business Bureau (www.bbb.org), Internet Crime Complaint Center (www.ic3.gov), and the National Consumers League (NCL) Fraud Center (www.fraud.org).

Above all, don't provide any additional information or money to the scammers. As soon as you suspect you've been had, halt all contact and cut off all access to your bank and credit card accounts. Sometimes the best you can hope for is to minimize your losses.

Protecting Against Computer Viruses and Other Malware

Any malicious software installed on your computer is dubbed *malware*. The two primary types of malware are *computer viruses* and *spyware*.

A computer virus is a malicious software program designed to do damage to your computer system by deleting files or even taking over your PC to launch attacks on other systems. A virus attacks your computer when you launch an infected software program, launching a "payload" that oftentimes is catastrophic.

Even more pernicious than computer viruses is the proliferation of spyware. A spyware program installs itself on your computer and then surreptitiously sends information about the way you use your PC to some interested third party. Spyware typically gets installed in the background when you're installing another program and is almost as bad as being infected with a computer virus. Some spyware programs will even hijack your computer and launch pop-up windows and advertisements when you visit certain web pages. If there's spyware on your computer, you definitely want to get rid of it.

Protecting Against Malware

You can do several things to avoid having your PC infected with malware. It's all about smart and safe computing.

- Don't open email attachments or files sent via text from people you don't know—or even from people you do know if you aren't expecting them. That's because some malware can hijack the address book on an infected PC, thus sending out infected email that the owner isn't even aware of. Just looking at an email message won't harm anything; the damage comes when you open a file attached to the email.

- Download files only from reliable file archive websites, such as Download.com (download.cnet.com) and Softpedia (www.softpedia.com). Do not download files you find on sites you don't know.

- Don't access or download files from music and video file-sharing networks, which are notoriously virus and spyware ridden. Instead, download music and movies from legitimate sites, such as the Amazon MP3 Store and the iTunes Store.

- Because viruses and spyware can also be transmitted via physical storage media, share USB drives, CDs, DVDs, and files only with users you know and trust.
- Use antimalware software, such as Windows Security, to identify and remove viruses and spyware from your system.

Using Antimalware Software

Windows 11 comes with its own antivirus utility built in. It's called Windows Security, and it tells you the last time your system was scanned, whether any threats have been detected, and more security-related information.

Of course, you're not locked into using Microsoft's antimalware solution. Several third-party programs are available, including the following:

- AVG Internet Security (www.avg.com)
- Avira Antivirus (www.avira.com)
- Bitdefender Antimalware Plus (www.bitdefender.com)
- IObit Malware Fighter (www.iobit.com)
- Malwarebytes for Windows (www.malwarebytes.com)
- McAfee Total Protection (www.mcafee.com)
- Norton 360 (us.norton.com)
- Trend Micro Antivirus+Security (shop.trendmicro.com)

If you just purchased a new PC, it might come with a trial version of one of these third-party antivirus programs preinstalled. That's fine, but know that you'll be nagged to pay for the full version after the 90-day trial. You can do this if you want, but you don't need to; remember, you have Windows Security built into Windows, and it's both free and very effective.

Delete Trial Versions

If you want to delete the trial version of any antivirus (or other) program, open the Settings app, click Apps, then click Apps & Features. Click the three-dot icon to the right of the program you want to get rid of and then click Uninstall. This will get rid of all the nagging from the program in question.

Whichever antimalware solution you employ, make sure you update it on a regular basis. These updates include information on the very latest viruses and spyware, and they're invaluable for protecting your system from new threats. (Windows Security is configured to update itself automatically—so there's nothing you have to do manually.)

It's Not All Good

Kaspersky Lab

When working with antimalware tools, you might encounter software from a company called Kaspersky Lab. This firm is based in Russia and, although it has a long history of use worldwide, it has been accused of contributing to the alleged Russian interference in our country's 2016 elections. The charges are serious enough that the U.S. Department of Homeland Security has banned Kaspersky products from all government departments.

As such, I can no longer recommend using Kaspersky's antimalware products. If you have any Kaspersky utilities installed on your computer, I recommend uninstalling them and switching to tools from a different vendor.

Using Windows Security

The easiest way to protect your computer from malware is with Windows 11's built-in Windows Security tool. In most instances, you don't have to access the tool at all; it's enabled and configured automatically. You can, however, view your security settings (and change any settings you want) from the Windows Security tool.

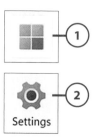

Settings

1. Click or tap the Start button to display the Start menu.

2. Click or tap Settings to display the Settings app.

3 Click or tap Privacy & Security.

4 Click or tap Windows Security.

5 Click or tap Open Windows Security to open the Windows Security tool.

6 The Home tab is selected by default. You see that your PC is being protected and when the last scan occurred. Any issues you need to address are highlighted here.

7 Click Virus & Threat Protection to change Windows Security settings.

8 Click Account Protection to configure security for your Microsoft account.

9 Click Firewall & Network Protection to configure Windows Firewall settings.

10 Click App & Browser Control to configure SmartScreen protection for the Edge browser and Windows apps.

11 Click Device Security to view device status and manage security for your computer hardware.

12 Click Device Performance & Health to view information about the health of your computer.

13 Scroll down and click Family Options to manage how your family uses their connected devices.

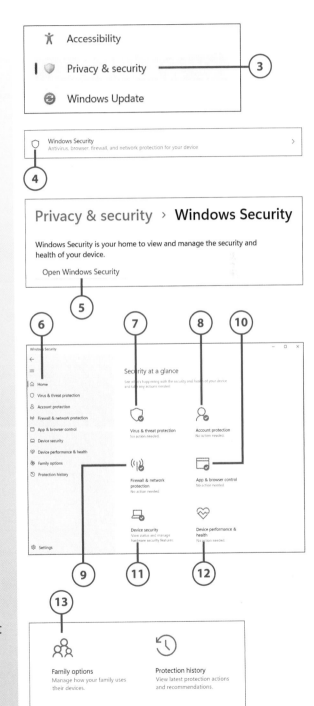

Protecting Against Ransomware

There's a type of malware that takes your computer hostage and won't let you access your files until you pay the hacker a monetary ransom. While this *ransomware* is typically targeted at large institutions (that can afford paying a large ransom), it sometimes hits individual PCs, with devastating results.

Fortunately, Windows 11 offers protection from ransomware, via what Microsoft calls controlled folder access. Activating this option keeps unauthorized applications, including ransomware, from accessing your computer's files and folders. You need to enable this option manually because it's not automatically enabled in Windows 11.

1. From the Windows Security tool, click or tap Virus & Threat Protection.

2. Go to the Ransomware Protection section and click or tap Manage Ransomware Protection.

3. Scroll down to the Controlled Folder Access section and click or tap "on" the Controlled Folder Access switch.

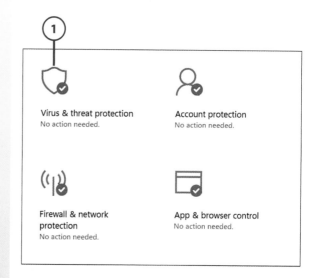

Virus & threat protection
No action needed.

Account protection
No action needed.

Firewall & network protection
No action needed.

App & browser control
No action needed.

Ransomware protection

No action needed.

Manage ransomware protection

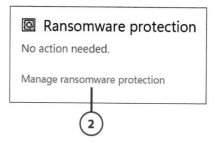

Controlled folder access

Protect files, folders, and memory areas on your device from unauthorized changes by unfriendly applications.

Off

>>>*Go Further*

ARE UPDATES LEGIT?

From time to time, you will inevitably be pestered to "update" something on your computer. This might be an update to Windows itself or to one of the programs you have installed.

Should you click "yes" when asked to install one of these updates? Or is this just another way to install malware on your system?

Although most updates are legitimate and necessary (they typically contain important bug fixes), some are just another way for the bad guys to install bad stuff on your system. This is especially so if the "update" notice is for a program or service you've never heard about and don't even have installed on your PC.

That said, if it's Windows that's asking you to approve the update, you should do it. Microsoft beams out updates to Windows over the Internet on a regular basis, and these *patches* (as they're called) help to keep your system in tip-top running condition. The same thing with updates to legitimate software programs; these updates sometimes add new functionality to the apps you use on a day-to-day basis.

So here's the rule: If it's an update to Windows or a program that you know and use on a regular basis, approve it. If it's an update to a program you don't use or don't know, then don't approve it. When in doubt, play it safe.

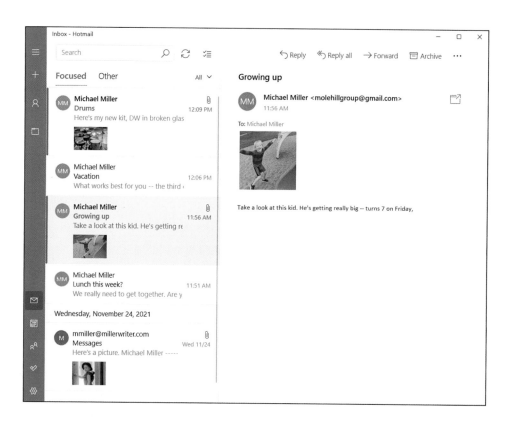

14

Emailing Friends and Family

When it comes to keeping in touch with the people you know and love, the easiest way to do so is via *email* (short for "electronic mail"). An email message is like a regular letter, except that it's composed electronically and delivered almost immediately via the Internet. You can use email to send both text messages and computer files (such as digital photos) to pretty much anyone with an Internet connection.

One of the easiest ways to send and receive email from your new PC is to use the Windows Mail app. You can also send and receive email in your web browser, using a web-based email service such as Gmail or Yahoo! Mail. Either approach is good and lets you create, send, and read email messages from all your friends and family.

Using the Windows Mail App

Windows 11 includes a built-in Mail app for sending and receiving email messages. By default, the Mail app manages email from any Microsoft email service linked to your Microsoft account, including Outlook.com and the older Hotmail. This means you'll see Outlook and

Hotmail messages in your Mail Inbox and will be able to easily send emails from your Hotmail or Outlook account.

Set Up Your Email Account

By default, the Mail app sends and receives messages from the email account associated with your Microsoft account. You can, however, configure Mail to work with other email accounts, if you have them. You launch the Mail app from the Windows Start menu.

Account Types

The Mail app lets you add Google (Gmail), iCloud (Apple), Office 365 (and Exchange), Outlook.com (as well as other Microsoft services, including Hotmail, Live.com, and MSN addresses), and Yahoo! accounts. You can also set up other email accounts, such as those from your Internet Service Provider or employer, using either POP or IMAP.

(1) From within the Mail app, click or tap the Settings button to display the Settings pane.

(2) Click or tap Manage Accounts to display the Manage Accounts pane.

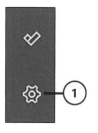

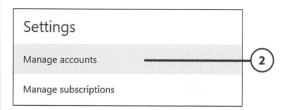

Settings

Manage accounts

Manage subscriptions

③ Click or tap Add Account to display the Choose an Account window.

④ Click or tap the type of account you want to add.

⑤ Enter the requested information and follow the onscreen instructions to complete the process. (This will be different for different services but typically includes your email address and password.)

Switching Accounts

To view the Inbox of another email account, click the name of that account in the Accounts section of the folders pane.

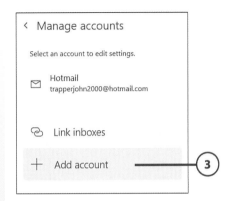

View Incoming Messages

All email messages sent to you from others are stored in the Inbox of the Mail app. Unread messages are displayed in bold.

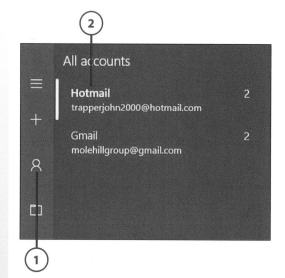

Resize the Window

By default, the Mail app launches in a squarish window. In this configuration, you only see two panes; when you click a message in the message pane, the content of that message then replaces the message pane. If you resize the window so that it's wider, or simply click the Maximize button to display the window full-screen, you'll see three panes, with a new content pane to the right of the message pane. In this configuration, when you click a message in the message pane, its contents display automatically in the content pane.

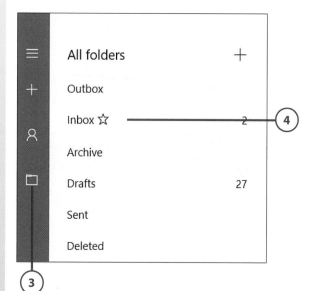

1 In the folder pane, click or tap the Accounts icon.

2 Click or tap to select the email account you want to use. (If you have multiple accounts, that is.)

3 In the folder pane, click or tap the Folder icon.

4 Click or tap the Inbox folder to display all messages in your inbox.

5 Click or tap the message you want to view; the contents of that message display in the contents pane.

6 If the message has a photo attached, click or tap the thumbnail to view the photo at a larger size in the Photos app. (To download a photo or other file to your computer, right-click or right-tap the item and select Save.)

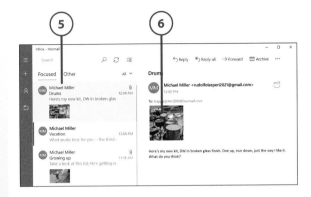

It's Not All Good

Download Danger

Be cautious when downloading and opening files attached to email messages. This type of file attachment is how computer viruses are often spread; opening a file that contains a virus automatically infects your computer.

You should only download attachments that you're expecting from people you know. Never open an attachment from a stranger. Never open an attachment you're not expecting. When in doubt, just ignore the attachment. That's the safest way to proceed.

Reply to a Message

Replying to an email message is as easy as clicking a button and typing your reply.

1 From an open message, click or tap Reply at the top of screen. The contents change to a reply screen, with the sender's email address already added to the To field.

Reply All

If the original message was sent to multiple recipients (including you), you also have the Reply All option, which sends your reply to everyone who received the original message. Don't click or tap the Reply All option by mistake if you want to reply only to the original sender!

2 Enter your reply at the top of the message; the bottom of the message "quotes" the original message.

3 Click or tap Send when you're ready to send the message.

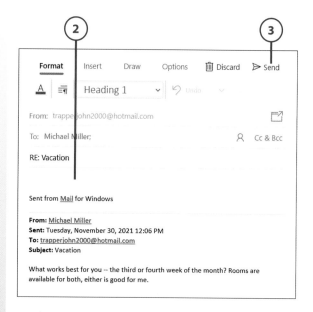

Send a New Message

Composing a new message is similar to replying to a message. The big difference is that you have to manually enter the recipient's email address.

1 Click or tap + New Mail at the top of the folders pane to display the new message.

2 Click or tap within the To field and begin entering the name or email address of the message's recipient.

3 If the name you type matches any in your contact list, Mail displays those names; select the person you want to email. (If there are no matches, continue entering the person's email address manually.)

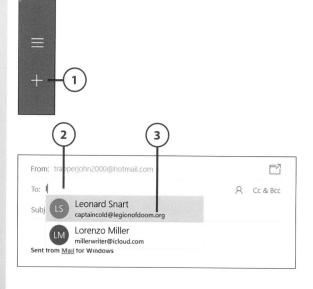

4 Click or tap the Subject field and type a subject for this message.

5 Click or tap within the main body of the message area and type your message.

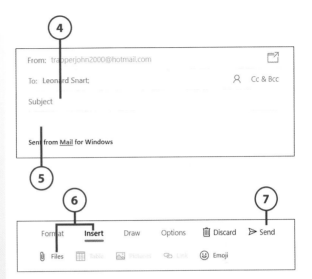

Formatting Your Message
Click the Format tab to apply Bold, Italic, and other formatting to your message text.

6 To send a file, such as a digital photo, along with this message, click or tap the Insert tab, click or tap Files, and then select the file.

7 Click Send to send the message and its attachment.

>>>Go Further
COPYING OTHER RECIPIENTS

You can also send carbon copies (Cc) and "blind" carbon copies (Bcc) to additional recipients. Just click Cc & Bcc at the right of the To field to display the Cc and Bcc fields; then enter the appropriate email addresses.

A Bcc differs from a Cc in that the Bcc recipients remain invisible to the other recipients. Use Cc when you want everyone to see everybody else; use Bcc when you want to keep those recipients private.

It's Not All Good

Don't Insert Pictures

The Insert tab includes an option to insert a picture (the Pictures button). In spite of the compelling name, do *not* use this option to insert a picture. This option places the picture, full-size, in line with the text in your document. This is confusing for recipients, makes it difficult to view the picture and read the accompanying text, and makes it more difficult for recipients to download the picture if they want. The better approach is to use the Attach Files option, which attaches the picture as a file to your message.

It's Not All Good

Large Files

Be careful when sending extra-large files (20MB or more) via email. Files of this size, such as large digital photos or home videos, can take a long time to upload—and just as long for the recipient to download when received. In addition, files that are too large may not be sent at all; many email providers have limits on the size of files that can be attached to email messages.

Move a Message to Another Folder

New messages are stored in the Mail app's Inbox, which is actually a folder. Mail uses other folders, too; there are folders for Outbox (messages waiting to be sent), Drafts, Junk (spam), Sent, Stored Messages, and Trash. For better organization, you can easily move messages from one folder to another.

(1) From within the messages pane, right-click or right-tap the message you want to move; then click or tap Move to display the Move To pane.

(2) Click or tap the destination folder (where you want to move the message). The message is moved to that folder.

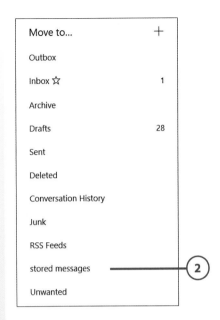

Delete a Message

There are several ways to delete a message in the Mail app.

(1) From within an open message, click or tap the Actions (three-dot) button; then select Delete. *Or...*

(2) Select the message in the message pane and then click or tap the Delete This Item (trash can) icon. *Or...*

(3) Select the message in the message pane and then press the Delete key on your computer keyboard. (Not shown.)

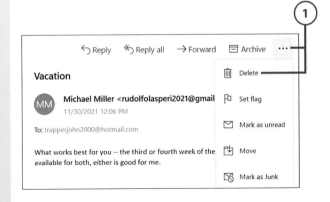

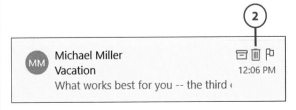

>>>Go Further

ACCESS YOUR EMAIL FROM ANYWHERE

If you have a Microsoft email account (with an @outlook.com, @live.com, or @hotmail.com address), you can check and send email from any computer, smartphone, or tablet, even when you're at work or away from home. All you have to do is use your device's web browser to go to the Outlook website at **www.outlook.com**.

From the site's main page, click Sign In and then enter your email address and password. You'll see all the messages in your inbox and other folders. As in the Mail app, click a message header to view the message contents, and click New Message to create and send a new message. You can sign in from anywhere!

Using Gmail

In addition to the email account you were given when you signed up for your home Internet service, you can add other email accounts you might have with various web-based email services. These services, such as Gmail and Yahoo! Mail, let you send and receive email from any computer connected to the Internet, via your web browser. They're ideal if you travel a lot or maintain two homes in different locations. (Snowbirds rejoice!)

Receive and Reply to Messages

Google's Gmail is the most popular web-based email service today. You can use any web browser to access your Gmail account, and send and receive email from any connected computer, tablet, or smartphone.

To sign up for a new account (it's free), use your web browser to go to mail.google.com, where you can set up a new account with an email address and a password. You can then send and receive email from any computer, just by signing in to your Google account.

1. Gmail organizes your email into types, each with its own tab: Primary, Social (messages from Facebook and other social networks), and Promotions (advertising messages). Most of your important messages will be the Primary tab, so click or tap that or another tab you want to view.

2. Click or tap the Inbox tab on the left to display all incoming messages.

3. Click the header for the message you want to view.

4. To download an attached folder or file, mouse over the item and then click or tap either the Download or Add to Drive icon. (Download saves to your computer; Add to Drive saves the file to Google Drive, Google's online storage service.) Alternatively, click Save to Photos to save a picture to Google Photos, Google's online photo service.

5. Reply to an open message by clicking or tapping Reply.

6. Enter your reply text in the message window.

7. Click or tap Send when done.

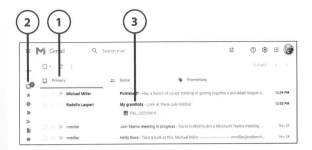

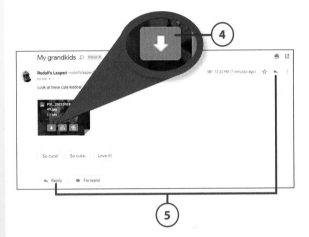

Send a New Message

New messages you send are composed in a New Message pane that appears at the bottom-right corner of the Gmail window.

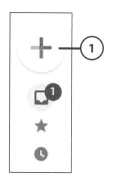

1 Click or tap Compose from any Gmail page to display the New Message pane.

2 Enter the email address of the recipient(s) in the To box. (Google might offer some suggestions, based on your contacts list and previous activity; click or tap a name to select it.)

3 Enter a subject in the Subject box.

4 Move your cursor to the main message area and type your message.

5 Attach a file to a message by clicking or tapping the Attach Files (paper clip) icon, navigating to and selecting the file you want to attach, and then clicking or tapping the Open button.

6 Send the message by clicking or tapping the Send button.

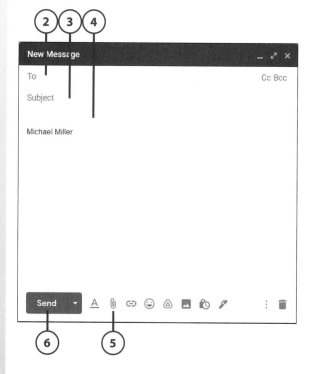

Delete a Message

You can easily delete messages you no longer need.

1 From an open message, click or tap the Delete icon in the message toolbar. Or...

2) From the messages pane, mouse over the message and then click or tap the Delete icon. *Or...*

3) From the messages pane, check to select those messages you want to delete.

4) Click or tap the Delete icon in the toolbar.

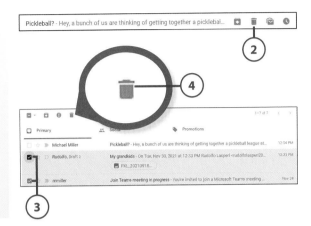

Managing Your Contacts with the People App

The people you email regularly are known as *contacts*. When someone is in your contacts list, it's easy to send her an email; all you have to do is pick her name from the list instead of entering her email address manually.

In Windows 11, all your contacts are managed from the People app, which is accessible from the Mail app. The People app connects to the Microsoft account you used to create your Windows account so that all the contacts from your main email account are automatically added. It can also connect to your other email accounts, including Gmail and Outlook.com. The People app serves as the central hub for everyone you interact with online.

First-Time Use

The first time you launch the People app, you may be prompted to let the app access and send email from your Microsoft account. You should answer in the affirmative and supply your email address and password if asked. You can later add other email accounts to the app.

View Your Contacts

The People app centralizes all your contacts in one place, and it even combines a person's information from multiple sources. So if a given person is a Facebook friend and is also in your email contacts list, his Facebook information and his email address appear in the People app. Launch the People app from the Windows Mail app.

(1) From within the Mail app, click or tap the People icon to open the People app.

(2) Click or tap a person's name to view that person's contact information.

(3) Click or tap the Email icon to send this person an email in the Windows Mail app.

(4) Click or tap the Map icon (next to an address) to view where this person lives.

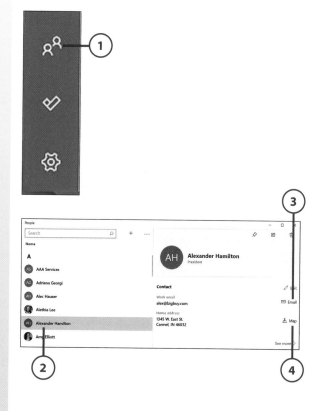

Add a New Contact

You can easily add new contacts within the People app.

(1) From within the People app, click or tap the + to display the New Contact pane.

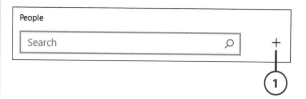

(2) Enter the person's full name into the Name box.

(3) Optionally, enter the person's mobile phone number into the Mobile Phone box.

(4) Optionally, enter the person's email address into the Personal Email box.

(5) To include additional email addresses, phone numbers, street addresses, or other information for this person, click + Email, + Phone, + Address, or + Other and enter the necessary information.

(6) Click or tap Save when done.

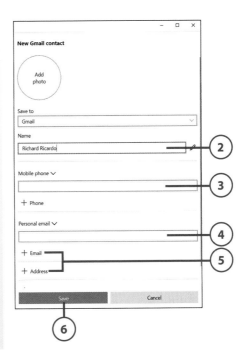

>>>Go Further

OPTIONAL AND ADDITIONAL INFORMATION

Many of the fields available when you create a new contact are optional. For example, you don't have to enter a person's company information if you don't want to.

You can also add more information than is first apparent. For example, you can enter additional phone numbers (for work, home, mobile, and the like) by clicking the + under the Phone box. It's the same thing if you want to enter additional email addresses, street addresses, companies, and other information; just click the appropriate + sign and enter the necessary information.

In this chapter, you discover how to use
Microsoft Teams and Zoom to talk one-on-one
with friends and family members.

→ Participating in Microsoft Teams Meetings
→ Participating in Zoom Meetings

Video Chatting with Microsoft Teams and Zoom

Your Windows 11 computer is a versatile communications device. Not only can you use your PC to send and receive emails, as I covered in Chapter 14, but you can also use your computer to make voice and video calls to other computer users using Microsoft Teams, Zoom, and other video chat services. All you need is your computer, a webcam and microphone (built into most laptop and all-in-one PCs), and a reliable Internet connection.

Participating in Microsoft Teams Meetings

You may often find yourself far away from the people you love. But that doesn't mean that you can't stay in touch—on a face-to-face basis.

When you want to talk to other people, use a video chat service that lets you connect to friends, family, and business associates in real time over the Internet. One such service, Microsoft Teams, is built into Windows 11 and automatically ties into your Microsoft Account. (There's even a Teams icon on the Windows taskbar by default.)

Unlike some other video chat services, Microsoft Teams is a full-featured collaborative communication platform, originally targeted at businesses, that includes a robust video meeting component. During the COVID-19 crisis, Microsoft revamped the video chat in Teams to be more user-friendly and appealing to individual consumers.

Microsoft Teams is part of the Microsoft 365 suite of applications. Microsoft offers it to businesses and other organizations on a subscription basis, ranging from $5 to $20 per user per month. (That's typically how this type of business software is marketed.) However, Microsoft also offers a free version of Teams with a limited feature set that's targeted at individual users. This free version focuses on video calling but also offers text chat and real-time collaboration via Microsoft Office apps.

Other Video Chat Services

Microsoft Teams and Zoom, discussed later in this chapter, are just two of several consumer-oriented video-calling services you can use on your Windows 11 PC. Other popular services include Facebook Messenger (for Facebook members only), Google Meet (meet.google.com), Skype (www.skype.com), and, if you have Apple computers and devices, FaceTime (support.apple.com/facetime).

Accept a Meeting Invitation

In Microsoft Teams, a video chat is called a meeting. A Teams meeting can include up to 300 people. When someone invites you to join a Teams meeting, you receive an invitation within Windows or via email.

(1) Click or tap Accept in the invitation.

2 If you're prompted to turn on your camera and microphone, do so.

3 Click or tap Join Now.

4 You are placed in a waiting room (called the Lobby) until the host admits you to the meeting.

5 When you join the meeting, you see the current speaker onscreen and your image in a live thumbnail. If there are more than just the two of you in the meeting, you see the other speakers in a grid.

6 Click or tap the Mute button to mute your microphone. Click or tap the button again to unmute your mic.

7 Click or tap the Turn Camera Off button to turn off your camera. Click or tap this button again to turn your camera back on.

8 Click or tap the Leave button to leave the meeting.

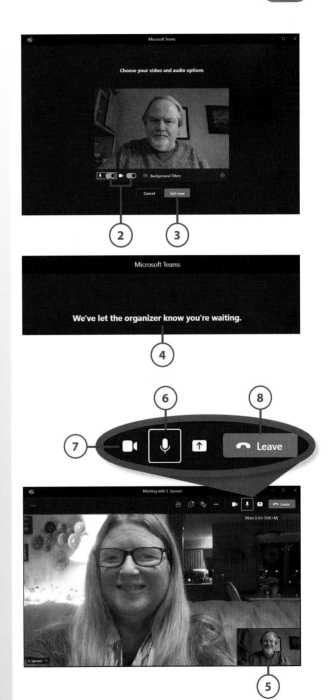

Blur Your Background

Microsoft Teams lets you blur the background behind you on the screen. This is good if don't want to show the other participants a messy room behind you or just want to make things look a little more interesting.

(1) From within a meeting, click or tap the More Actions (three-dot) button.

(2) Select Apply Background Effects to open the Background Settings panel.

(3) Click or tap Blur.

(4) Click or tap Preview to see what the effect looks like.

(5) Click or tap Apply to apply the effect.

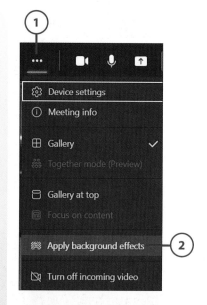

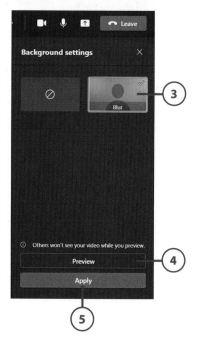

6 You appear onscreen with a blurred background.

7 Click the X to close the Background Settings panel.

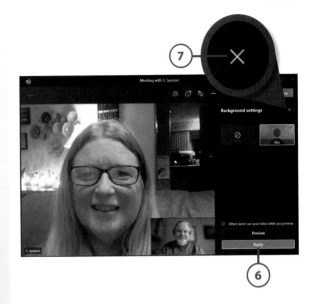

Launch a New Teams Meeting

When you want to host your own Microsoft Teams meeting, you can invite any of your contacts in the People app or other people via their email addresses.

1 Click or tap the Chat icon in the Windows taskbar to display the Teams panel.

2 Recent activity is listed here. Click or tap one of these to resume that meeting or text chat. *Or…*

3 Click or tap Meet to start a new meeting.

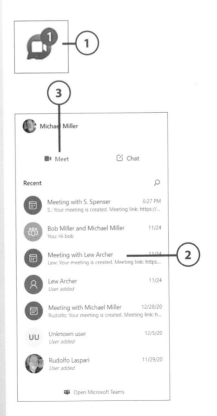

(4) Make sure your camera and microphone are turned on.

(5) Click or tap Background Filters if you want to blur your onscreen background.

(6) Click or tap Join Now to start the meeting.

(7) You're prompted to invite other people to the meeting. Click or tap Share via Default Email.

Send a Link

You can also paste a link to the meeting into other mail or messaging programs or social media messages. Just click or tap Copy Meeting Link and then paste that link into the other app.

(8) Your default email app opens with a new message created. Enter the email addresses of the people you want to invite to the meeting and then send the message.

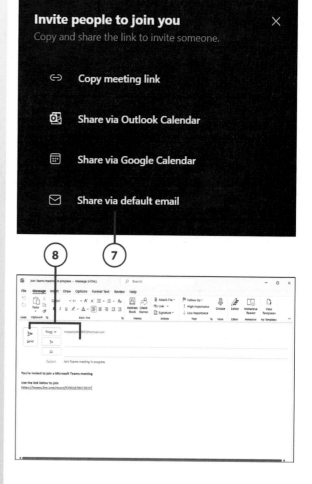

(9) Close the Invite People to Join You window and return to the main meeting window.

(10) When a recipient clicks the link in the invitation you sent, they're placed in the virtual Lobby, and you see an onscreen message. Click or tap Admit to let them into the meeting.

(11) You see the other people in your meeting. Your live picture appears in a thumbnail in the corner.

(12) The other participants are listed in the Participants panel on the right side of the window. Click or tap Share Invite to invite other people to the meeting.

(13) Click or tap the Leave button to leave the meeting but leave it up and running for other participants. Or...

(14) Click or tap the down arrow next to the Leave button and then click or tap End Meeting to end the meeting for all participants.

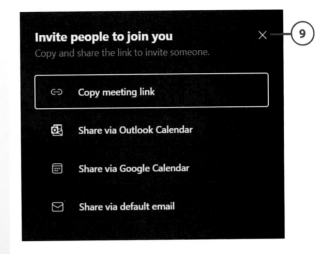

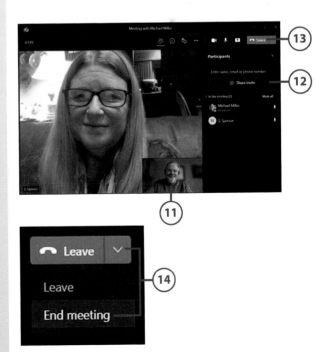

(15) When prompted to end the
meeting, click or tap End.

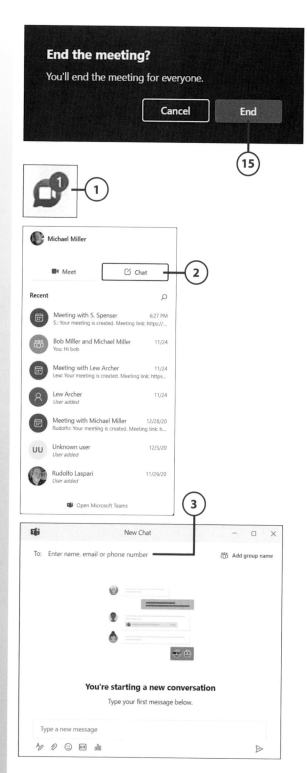

End the meeting?

You'll end the meeting for everyone.

Cancel   End

(15)

Start a Text Chat

Microsoft Teams isn't just about video
meetings; it also lets you participate in
text chats with your contacts.

(1) Click or tap the Chat button on
the taskbar to open the Microsoft
Teams panel.

(2) Click or tap the Chat button.

(3) Enter the name of a contact or
that person's phone number or
email address into the To field.

Group Chats

You can enter multiple names to initiate a
group text chat.

4 Enter your message into the Type a New Message field and then press Enter. Your message is sent.

5 Your messages appear on the right side of the window. Messages from other participants appear on the left side of the window.

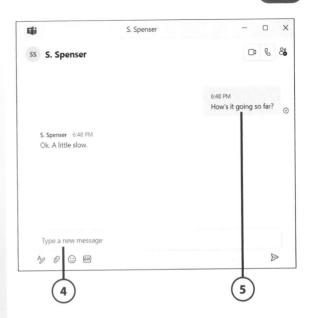

My Video Chat for Seniors

Learn more about Microsoft Teams, Zoom, and other video chat services in my companion book for AARP, *My Video Chat for Seniors*. It's available wherever good books are sold.

Participating in Zoom Meetings

Microsoft Teams is just one of several services offering online video chat. Arguably the most popular video chat service is Zoom, which zoomed in popularity during the first few months of the COVID-19 pandemic. When everyone was stuck at home and couldn't go to the office or to school, businesses used Zoom to connect their employees for virtual meetings and collaboration; schools used Zoom to conduct distance learning; and individuals used Zoom just to talk to one another.

In Zoom world, a video chat is called a Zoom meeting. You can have up to 100 participants in a Zoom meeting, and each chat can last up to 40 minutes long. (If you need more time, you can easily launch or schedule a second chat immediately following the first one.)

Zoom runs as a free app on your computer, and the basic service is also free. The first time you use Zoom (from Zoom's website), you're prompted to download and install the app. You can also install it before you start using Zoom, from Zoom's website, if you want. Learn more (and download the app) at www.zoom.us.

>>>Go Further
ZOOM FOR BUSINESS

Zoom started as a video conferencing service for large enterprises and continues to serve that market. Zoom for business differs from the home-based version you know by allowing longer meetings with more participants, cloud-based recording, user management via an admin portal, and detailed reporting.

Unlike Zoom's home version, the business version of Zoom isn't free. Zoom offers several different plans, starting at $14.99 per month and going up from there. Zoom also offers several business add-ons to its basic service, including options for audio calling, cloud storage, and larger groups.

If you or someone you know is interested in Zoom for business, go to www.zoom.us/pricing for more information.

Accept a Meeting Invitation

When someone else is hosting a Zoom meeting, that person sends out invitations, typically via email, to all participants. This is true of both instant meetings (those being held immediately) and those scheduled for a future time.

(1) From within the email, click the link for the Zoom meeting.

First Meeting

If this is your very first Zoom meeting and you haven't installed the Zoom app, you'll be prompted to download the Zoom app to your computer. If you've already downloaded the app but not signed in, you'll be prompted to do that now.

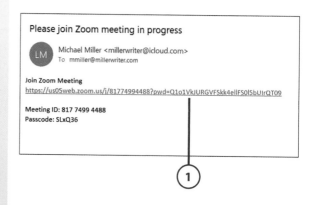

Please join Zoom meeting in progress

LM   Michael Miller <millerwriter@icloud.com>
     To mmiller@millerwriter.com

Join Zoom Meeting
https://us05web.zoom.us/j/81774994488?pwd=Q1o1VkJURGVFSkk4ellFS0l5bUIrQT09

Meeting ID: 817 7499 4488
Passcode: SLxQ36

2 If you're prompted to join with your device's audio and/or video, do so.

3 You may be placed in a virtual waiting room until admitted by the meeting leader. This is particularly the case if you join a scheduled meeting a few minutes early. Just sit back and cool your heels.

4 Once you're admitted to the meeting, you're ready to go. You see a large image of the meeting leader in the window and a smaller thumbnail of you. Other participants may appear in similar thumbnails, or you may all appear in a grid.

5 Click or mouse over the screen to display the chat controls.

6 Click the Mute Audio icon to mute your microphone. Click this icon again to unmute your mic.

7 Click the Stop Video icon to turn off your computer's camera. Click this icon again to turn your camera back on.

8 Click the red Leave button to leave the meeting.

Enter a Meeting Manually

Clicking or tapping a link is the easiest way to join a Zoom meeting, but it's not the only way. When you receive an invitation via email or text, that invitation should include the meeting ID and optional passcode that you can enter manually into the Zoom app. This is particularly useful if you receive a text invitation on your phone but want to Zoom using another device, such as your tablet or computer.

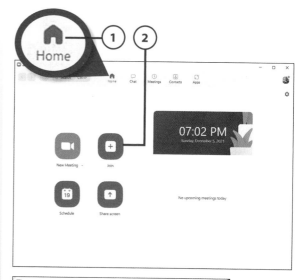

(1) When it's time for the meeting, launch the Zoom app on your computer. Sign in, if necessary, and then select the Home tab.

(2) Click or tap the Join icon.

(3) Enter the meeting ID into the Meeting ID field.

(4) Accept or change your name.

(5) Click or tap Join.

(6) If prompted for a passcode or password, enter it and then click or tap Join Meeting.

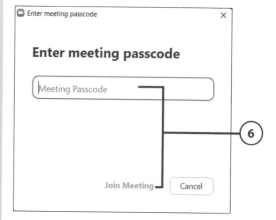

7 If prompted, click Join with Video.

8 If you're placed in a virtual waiting room, wait to be admitted.

9 Once you're admitted to the meeting, you're ready to go.

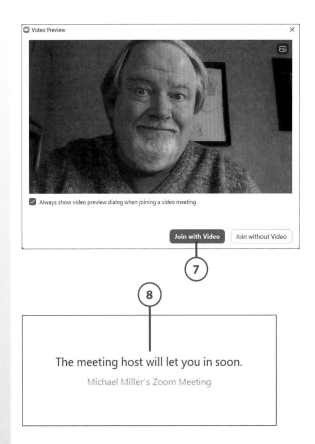

>>>Go Further
SWITCHING VIEWS

When you're in a Zoom meeting, you have the choice of viewing the other participants in one of two views. You switch between views by clicking the View button and selecting the view you want.

Speaker View puts the person currently speaking in the large video window, with up to three other participants in smaller thumbnails. In this view, the person in the large window is constantly changing, depending on who's talking.

Gallery View displays many participants (up to 49 at a time) in a grid layout. The person currently speaking is highlighted with a green border. If there are more participants than can fit on screen, you can scroll through additional participants by clicking the right or left arrows on your keyboard.

Apply a Virtual Background

The virtual background option is one of the most fun options Zoom offers, and it's the one I'm asked about the most. Instead of the other participants looking at the room or blank wall beyond you, you can add a virtual background that makes it appear as if you're somewhere else. It's easy to do—and you can choose from Zoom's stock backgrounds or any image stored on your computer. You can even download other backgrounds from the Internet!

Solid-Color Background

Zoom's virtual backgrounds work best if you're sitting in front of a solid-color background. They work even better if the background is green. (This is the fabled "green screen effect.") For best effect, you can set up an actual green screen, in the form of a green cloth or paper backdrop, which you can find online or at your local photography store.

1. During a Zoom meeting, click or tap the up arrow next to the Stop Video button.

2. Click or tap Choose Virtual Background. This displays the Settings window with the Virtual Background tab selected.

3. Select the Video Backgrounds tab. Zoom's built-in backgrounds are displayed. You'll also see any other backgrounds you've recently selected.

4. Click or tap one of these backgrounds to use it. (Click or tap Blur if you simply want a blurred background.) You see a preview of how the virtual background looks.

5. If you have a green screen background, check the I Have a Green Screen option.

6. Check Mirror My Video to view the virtual background as others see it.

7. Return to your normal background by selecting None.

8. Close the Settings window when you're done.

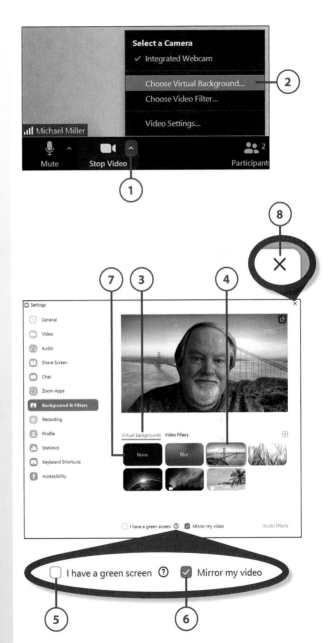

Leave a Meeting

When a meeting officially ends, you and all other participants are automatically disconnected from it. You can, however, leave a meeting before it officially ends, which is common.

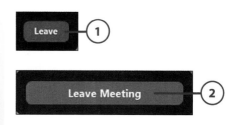

1. Tap or mouse over the screen to display the chat controls; then click or tap the red Leave icon.

2. Click or tap Leave Meeting.

Start a New Instant Meeting

An instant meeting is one that you start immediately. It's not scheduled in advance.

Anyone with a Zoom account can start an instant meeting. Once you've started one, you can then invite your participants.

1. From within the Zoom app, select the Home tab; then click or tap New Meeting.

2. If you're prompted to connect or use your device's audio and/or video, do so.

3. Your meeting is now live, with you as the only participant until you invite others to your meeting. Click or tap Participants to open the Participants panel.

4. Click or tap Invite.

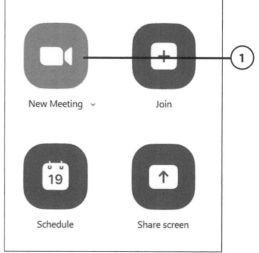

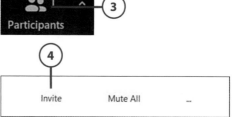

5. Click or tap to select the Email tab.

6. Select your email client. (In most cases, you should select Default Email.)

7. You see a new email message with the meeting information already entered. Enter the email address(es) of your desired participant(s) and click or tap to send the invitation.

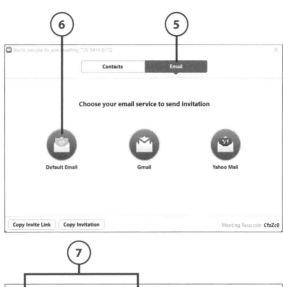

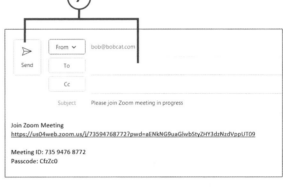

Schedule a Meeting in Advance

What if you want to Zoom with your grandkids on Thursday at 7:00 p.m. or have a regularly scheduled book club meeting over Zoom every Wednesday morning at 8:00 a.m.? Zoom lets you schedule meetings in advance so others can plan to attend.

1. From the Home tab, click or tap Schedule. The Schedule Meeting window opens.

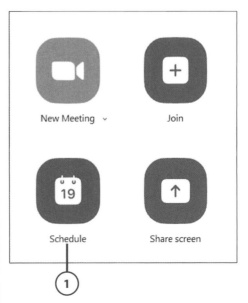

(2) Enter a name or topic for the meeting.

(3) Enter the start date and time.

(4) Enter the length or duration of the meeting—up to a maximum of 40 minutes on a free account.

(5) If it's a recurring meeting (one that happens on the same day every week or month, or the same time every day), click or tap Recurring Meeting and select how often it repeats.

(6) Select whether you want Zoom to generate an automatic meeting ID or use your personal meeting ID. (In most instances, let Zoom generate the ID automatically for better security.)

(7) Make sure Waiting Room is checked.

(8) Select whether you want the host video (your video) on or off. (You can change this during the meeting if you want.)

(9) Select whether you want participants' video on or off. (You can also change this during the meeting.)

(10) Select if you want this meeting added to a specific calendar app.

(11) Click Save.

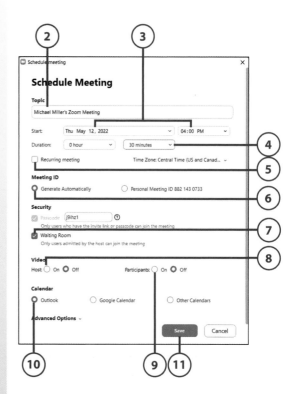

Starting and Ending a Meeting

There are a few tasks you need to undertake to both start and end a meeting.

(**1**) When you launch a new meeting, participants who log in are ushered into a virtual waiting room, where they stay until you admit them into the meeting. As each participant enters, click or tap Admit to admit each one into the meeting. (This is only if you have the waiting room option enabled; otherwise, participants are immediately admitted to the meeting.)

(**2**) End the meeting by clicking or tapping the red End button.

(**3**) Click or tap End Meeting for All.

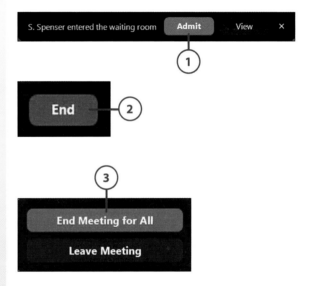

>>>Go Further

WEBCAMS

Most laptop and all-in-one PCs have webcams built in. You can use your laptop's built-in webcam to make video calls with Microsoft Teams, Zoom, and other video chat services. Because the webcam includes a built-in microphone, you can also use it to make voice calls.

If your PC doesn't have a built-in webcam, you can purchase and connect an external webcam to make Zoom and Microsoft Teams or Skype video calls. Webcams are manufactured and sold by Logitech and other companies, and they connect to your PC via USB. They're inexpensive (as low as $40 or so) and sit on top of your monitor. After you've connected it, just smile into the webcam and start talking.

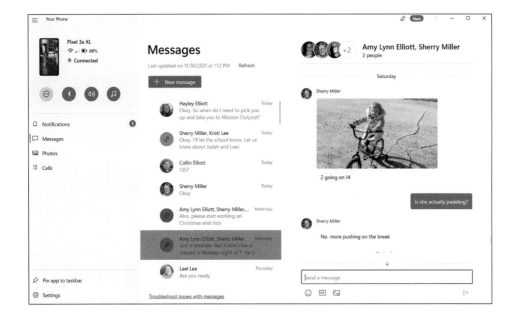

In this chapter, you learn how to link your Android smartphone to your Windows 11 PC to send and receive text messages, make phone calls, and more.

→ Linking Your Phone to Your PC
→ Texting and Calling on Your PC
→ Doing More with Your Phone and Windows

16

Using Your Windows PC with Your Android Phone

If you'd like to use your PC to send and receive text messages from your Android mobile phone, use the Your Phone app included with Windows 11. This app synchronizes your personal computer to your cell phone so you can send and receive text messages from your PC. (And it's a lot easier to type texts on your big computer keyboard than it is on the tiny one on your smartphone!) You can also share web pages and other documents between your phone and your PC.

It's Not All Good

Android Only

The Your Phone app and the phone/PC functionality in Windows 11 is designed to work specifically with Android phones. This includes phones from Google, Motorola, Nokia, OnePlus, Oppo, Samsung, Sony, and others.

The Your Phone app is not designed to work with Apple's iPhones. Although you technically can link an iPhone to your Windows 11 PC, it won't have the same functionality as an Android phone. (About the only thing you can do with an iPhone is send web page links from your phone to your PC; you can't text or call or do much of anything else from your PC to an iPhone.)

Microsoft blames this on Apple making it difficult for outside (non-Apple) developers to connect to its iOS phone operating system. Apple says it does this to ensure stronger security for its devices and, besides, you probably ought to be using a Mac instead of a Windows PC, anyway. (Admittedly, the Apple ecosystem makes it very easy to use Apple devices together.) Bottom line, if you have an iPhone, you can't use the Your Phone app. It's an Android-only thing.

Linking Your Phone to Your PC

For your Android phone and Windows 11 computer to share text messages and other data, you first have to install the Your Phone Companion app on your phone. You can find the Your Phone Companion app in your phone's app store; it's free.

Link Your Phone

Once you've installed the Your Phone Companion app on your Android phone, you need to link your Windows 11 computer to your phone. You do this by configuring both your computer and your phone.

1. On your computer, open the Your Phone app. If you do not yet have your phone linked, click the Get Started button.

2. If you haven't installed the Your Phone Companion app on your phone, do so now. When the app is installed on your phone, open the app and tap Link Your Phone and PC. (You may be prompted to sign in at this point; if you are so prompted, do it.)

3. Back on your computer, in the Your Phone app, check I Have the Your Phone Companion – Link to Windows App Ready.

4. Click Pair with QR Code.

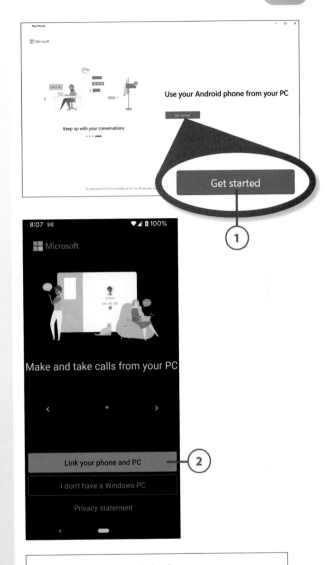

(5) On your phone, when asked if the QR code on your PC is ready, tap Continue.

(6) Follow the onscreen instructions to use your phone to scan the QR code displayed on your computer.

7 If the Your Phone Companion app on your phone prompts you to allow permissions for various activities, click Allow to do so.

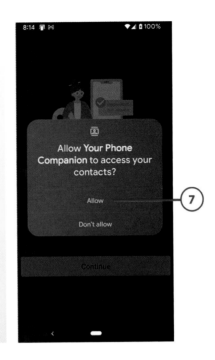

Texting and Calling on Your PC

When your phone and your PC are linked, you can use your computer to send and receive text messages—and to make phone calls.

Receive Text Messages

Once the Your Phone app is linked to your Android phone, you can access all the texts you receive on your phone on your computer, as well. This makes it easier for you to text when you're working with your computer.

1 Click or tap the Messages tab in the Your Phone app to view recent texts you've received. Unread texts are in bold.

2 Select the text to which you're replying.

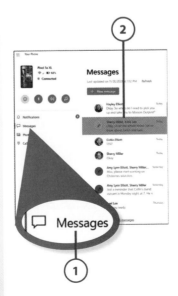

③ Enter your reply into the Enter a Message box.

④ Click or tap the Send icon or press Enter.

Send Text Messages

You can initiate new individual (but not group) texts from within the Your Phone app.

① Click or tap the Messages tab.

② Click or tap an existing conversation to resume that conversation. *Or…*

③ Click or tap the New Message button to start a new conversation.

④ Start typing the name or phone number of the person you want to text.

⑤ Matching names from your contact list are displayed. Click or tap to select the person you want to text.

⑥ Type your message into the Send a Message box.

⑦ Click or tap the Emoji button to insert an emoji.

⑧ Click or tap the GIF button to insert a GIF (animated picture).

⑨ Click or tap the Attach Image button to send a photo or other image file stored on your computer.

⑩ Click or tap the Send icon or press Enter on your keyboard to send the text.

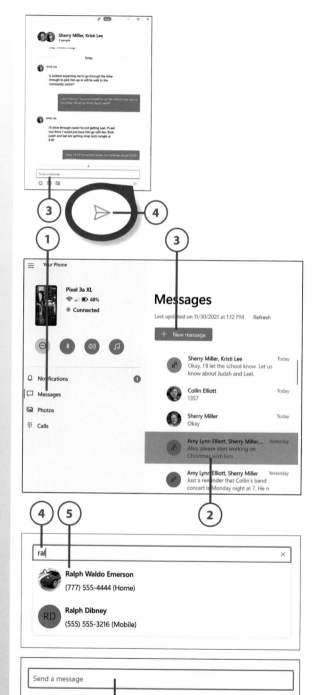

Make a Phone Call

When properly configured, the Your Phone app enables you to make calls on your Android phone from your Windows 11 computer.

1. Click or tap to display the Calls tab.

2. Call one of your contacts by entering that person's name into the Search Your Contacts box. *Or...*

3. Click or tap the keypad numbers to enter a phone number you want to call or use your computer keyboard to enter the number manually.

4. Click or tap the green Dial button or press Enter on your keyboard to place the call.

5. The Your PC app connects to your phone, initiates the call, and displays a Call panel. Click or tap the Down arrow to expand the panel.

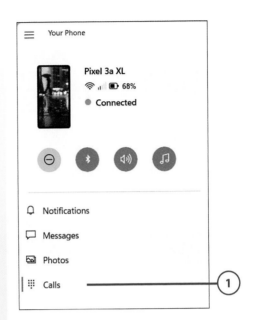

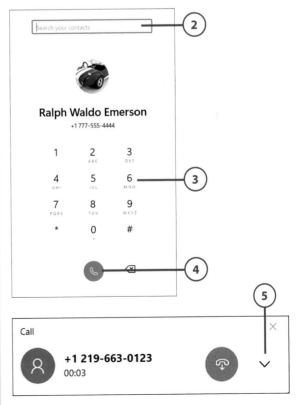

6 Click or tap the Mute button to mute the call.

7 Click or tap the Keypad button if you need to press any numbers during the call.

8 Click or tap the Use Phone button to transfer the call back to your phone.

9 Click or tap the red Disconnect button to end the call.

Doing More with Your Phone and Windows

When you've paired your Android phone with your Windows 11 PC, you can do more than just text or make phone calls. You can also share web pages between your devices and view your phone's photos on your PC.

Share a Web Page from Your Phone to Your PC

Ever find a web page when you're browsing on your phone and want to view it full-size on your computer screen? Windows 11 makes it easy to continue reading interesting web pages when you switch from your Android phone to your PC.

1 On your phone's web browser, navigate to the web page you want to share; then tap the Share button or link.

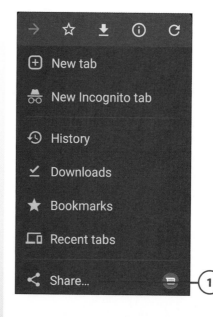

2 Select Your Phone.

3 Select the PC you want to link to. The Edge browser opens on your computer and displays the web page you were viewing on your phone.

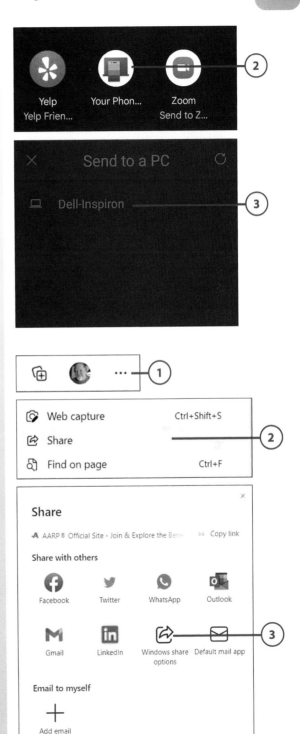

Share a Web Page from Your PC to Your Phone

Just as you can share a web page from your phone to your PC, you can also share a page from your PC to your Android phone.

1 On your computer, use the Microsoft Edge browser to navigate to the web page you want to share; then click or tap the Settings and More (three-dot) button to display the pull-down menu.

2 Select Share to open the Share window.

3 Click or tap Windows Share Options to open the Share Link window.

4 Click or tap Your Phone in the Share with App section.

5 You receive an alert on your phone from the Your Phone Companion app. Tap this alert, and your phone's web browser opens with the web page you shared displayed.

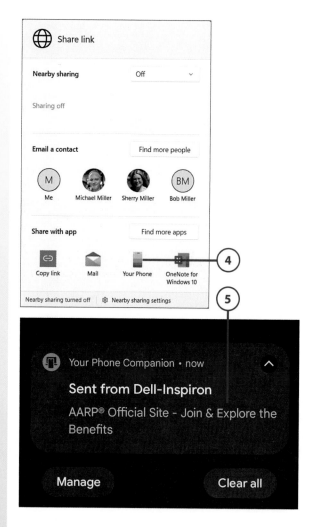

View Photos from Your Phone on Your PC

If you're like me, you take a lot of pictures with your smartphone. Now, with the Your Phone app, you can easily view all your phone pictures on your computer—and save them to your PC.

1 From within the Your Phone app on your PC, click to select the Photos tab.

2 All the photos on your phone are displayed here. Click or tap a picture to view it larger.

(**3**) Click Open to edit the selected photo in the Windows Photos app.

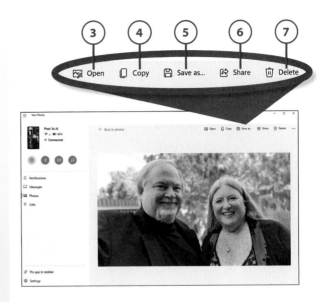

Photos App

Learn more about the Windows Photos app in Chapter 18, "Storing, Editing, and Sharing Your Pictures."

(**4**) Click or tap Copy to copy the photo so you can paste it into another application.

(**5**) Click or tap Save As to save the photo to a location on your computer.

(**6**) Click or tap Share to share the photo with another Windows app.

(**7**) Click or tap Delete to delete this photo from your phone.

>>>Go Further

RUNNING ANDROID APPS ON YOUR PC

Microsoft is poised to add a very useful feature to Windows 11 in the coming months. (It may even be live by the time you read this book!) This new functionality lets you run Android apps within Windows 11. These are the same apps that run on Android phones and tablets and should give you a lot more options for your Windows 11 PC.

Here's how it's supposed to work: You'll need to install the Amazon Appstore app on your computer. This app should be available in the Microsoft Store. Once it's installed, you'll need to sign in to or create a new Amazon account, and then you'll be able to browse the Amazon Appstore for Android apps. Download and install an app to use it on your PC just like you do on an Android phone. Look for this new functionality in the near future!

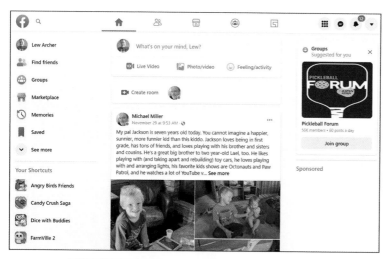

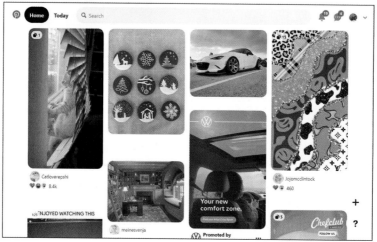

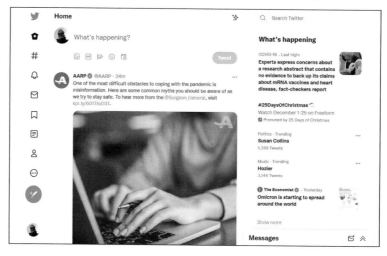

In this chapter, you find out how to use Facebook and other social networks to connect with friends and family.

→ Sharing with Friends and Family on Facebook
→ Pinning Items of Interest to Pinterest
→ Keeping Up to Date with Twitter

17

Connecting with Facebook and Other Social Media

When you want to keep track of what friends and family are up to and keep them up to date on your activities, Facebook can be a great way to do it. Facebook is a *social network*, which is a website that lets you easily share your activities with people you know. Write one post, and it's seen by hundreds of your online "friends." It's the easiest way I know to connect with almost everyone I know.

Facebook isn't the only social network on the Internet, however. Other social media, such as Pinterest and Twitter, target particular types of users. You might find yourself using social media other than Facebook to keep in touch with friends and family.

Sharing with Friends and Family on Facebook

A social network is a website community that enables users to connect with and share their thoughts and activities with one another. Think of it as an online network of friends and family, including former schoolmates, coworkers, and neighbors.

The largest and most popular social network today is Facebook, with close to 3 billion active users worldwide each month. Although Facebook started life as a social network for college students, it has since expanded its membership lists, and it is now the preferred social network for more mature users.

Signing Up for Facebook

To use Facebook, you have to sign up for an account and enter some personal information. Use your web browser to go to www.facebook.com, click Create New Account, and follow the onscreen instructions. It's free to sign up for and use Facebook

Discover New—and Old—Friends on Facebook

To connect with someone on Facebook, you must become mutual *friends*. A Facebook friend can be a real friend, or a family member, colleague, acquaintance—you name it. When you add people to your Facebook friends list, they see everything you post—and you see everything they post.

App or Website?

Although there is a dedicated Facebook app you can download from the Microsoft Store (there's probably an icon for it on your Start menu), most people access Facebook from their web browser at www.facebook.com. The interface of the app mimics the interface of the website, so either approach works.

The easiest way to find friends on Facebook is to search for a particular person.

(1) Click or tap the Search icon on the tool bar. This expands the icon into a Search box.

(2) Enter the person's name into the Search box and then press Enter.

(3) On the search results page, click the People option in the Filters pane.

(4) Fine-tune your search by using the controls in the Filters pane. For example, you can filter the results by city, education (schools attended), and work.

(5) If your friend is listed, click the Add Friend button to send that person a friend request.

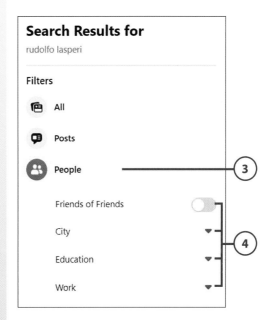

Friend Requests

Facebook doesn't automatically add a person to your friends list. Instead, that person receives an invitation to be your friend to accept or reject. To accept or reject any friend requests you've received, click the Friend Request button on the Facebook toolbar. (And don't worry; if you reject a request, that person won't be notified.)

Post a Status Update

To let your family and friends know what you've been up to, you need to post what Facebook calls a *status update*. Your status updates are broadcast to people on your friends list via the News Feed on their home pages. A basic status update is text only, but you can also include photos, videos, and links to other web pages in your posts.

1. Click or tap Home on the Facebook toolbar to return to your home page.

2. Click or tap within the Create Post ("What's on your mind?") box near the top of the page. The box expands to offer more options.

3. Type within the box to enter your message.

4. If you're with someone else and want to mention them in the post, click or tap Tag Friends and enter that person's name.

5. Include a picture or video with your post by clicking or tapping Photo/Video to display the Open dialog box; then select the photos or videos to include.

6. Adjust who can read this post by clicking or tapping the Sharing To button and making a selection.

7. Add a location or place to your post by clicking or tapping Check In and making a selection.

8. Post your update by clicking or tapping the Post button.

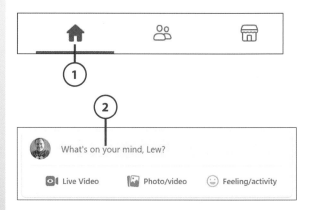

Who Sees Your Posts?

You can opt to make any post Public (meaning anyone can read it), visible only to your Friends, or visible to select people and groups.

View Posts in Your News Feed

Your home page on Facebook displays a News Feed of status updates made by people on your friends list. Posts that Facebook thinks you're interested in are at the top; scroll down through the list to read additional posts.

1. Click or tap Home on the Facebook toolbar to return to your home page.

2. Your friends' posts are displayed in the News Feed in the middle of the page. To leave your own comments about a post, click or tap Comment, enter your text into the resulting text box, and then click or tap Post.

3. To "like" a post, mouse over Like and select from one of the available emojis.

4. If a post includes a link to another web page, that link appears beneath the post, along with a brief description of the page. Click or tap the link to open the other page in your web browser.

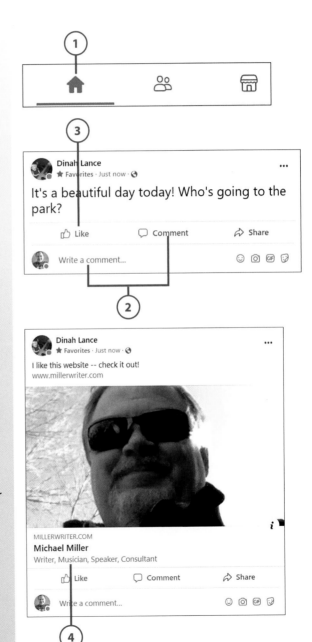

5 If a post includes one or more photos, click or tap the photo to view it in a larger onscreen lightbox.

6 If a post includes a video, playback should start automatically. (If not, click or tap the Play button.) Click or tap the Pause button to pause playback.

7 Click or tap the volume control to unmute the sound.

Pinning Items of Interest to Pinterest

Facebook isn't the only social network that might be of interest to you. Pinterest (www.pinterest.com) is a social network with particular appeal to older adults, especially women—although Pinterest has a lot of male users, too.

Unlike Facebook, which lets you post text-based status updates, Pinterest is all about images. The site consists of a collection of virtual online "boards" that people use to share pictures they find interesting. Users save or "pin" photos and other images to their personal boards, and then they share their pins with online friends.

You can pin images of anything—clothing, furniture, recipes, do-it-yourself projects, and the like. Your Pinterest friends can then "repin" your images to their boards—and on and on.

Joining Pinterest

Like other social media sites, Pinterest is free to join and use. You can join with your email address or by using your Facebook or Google account login.

Create a New Board

Pinterest lets you create any number of boards, each dedicated to specific topics. If you're into quilting, you can create a Quilting board; if you're into gardening, you can create a Gardening board with pictures of inspiring gardens.

1. From the top-right corner of the Pinterest home page (www.pinterest.com), click or tap your picture or profile icon to display your profile page.

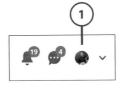

② Click the + icon.

③ Click Board to display the Create Board panel.

④ Enter the name for this board into the Name box.

⑤ Click or tap to select Keep This Board Secret if you don't want others to see this board.

⑥ Click or tap the Create button.

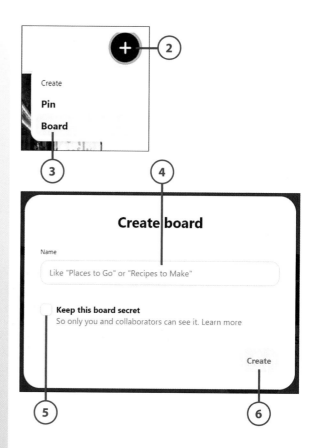

Find and Save Interesting Items

Some people say that Pinterest is a little like a refrigerator covered with magnets holding up tons of photos and drawings. You can find lots of interesting items pinned from other users—and then save them to your personal boards.

① Enter the name of something you're interested in into the Search box at the top of any Pinterest page and then press Enter. Pinterest displays pins that match your query.

② Mouse over the item you want to save to display the control panel.

③ Click or tap Save if you want to save to the suggested board.

④ Save to a different board by clicking the down arrow, selecting a board, and then clicking or tapping Save.

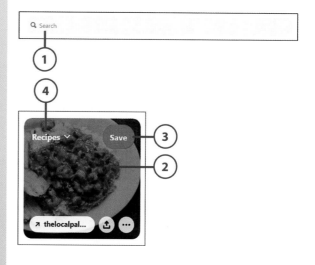

Save an Item from a Web Page

Pinning images from a web page is as easy as copying and pasting the page's web address.

It's Not All Good

Copyright

As noted on its website, "Pinterest is not the copyright holder in the images that users pin on the site. Where necessary, you should get permission to use an image from its copyright owner."

While you *should* get permission before pinning, practically no one ever does, or Pinterest would cease to exist. What copyright holders do, if they don't want a particular image shared on Pinterest, is submit a copyright removal request to the Pinterest site. If Pinterest agrees to that request, and it almost always does, it will remove the pinned image in question and notify the original poster of that removal.

1. From Pinterest's main page, click or tap the + button in the lower-right corner.

2. Click or tap Create a Pin.

3. Click or tap Save from Website.

4. Enter the web address (URL) of the page you want to pin into the Enter a Website Link box, and then press Enter.

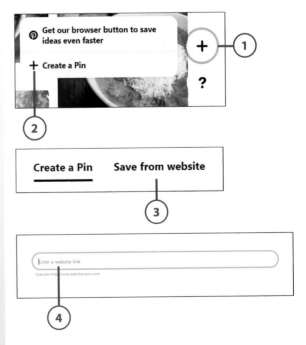

5 Pinterest displays all images found on the selected web page. Click or tap to select the image(s) you want to pin.

6 Click the Add Pin button.

7 Enter a title for this pin into the Add Your Title field.

8 Enter a description for this pin into the Tell Everyone What Your Pin Is About field.

9 Click or tap the Select down arrow and select the board to which you want to pin this image.

10 Click or tap Save. The item is now pinned to the selected board.

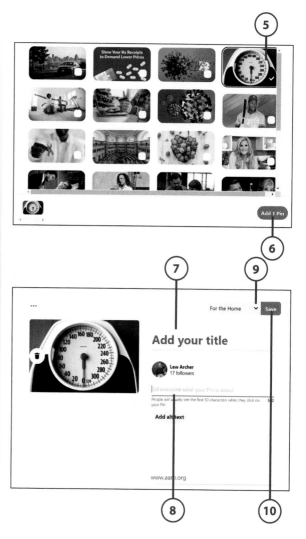

Pin It Button

Pinning a page is even easier if the page displays its own Pin It button. Just click the button, select an image from the page, and you're good to go.

Keeping Up to Date with Twitter

Twitter is like Facebook but without the group social interaction. The focus is on relatively short (280-character) messages called *tweets*, which are shared with a user's followers. In addition to text, a tweet can include photos, videos, and links to other websites.

Twitter is immensely popular with news outlets, celebrities, and politicians, many of whom use Twitter to communicate with their large numbers of followers. Most individuals use Twitter to follow others' tweets and retweet those posts they find most interesting.

Signing Up

Signing up for Twitter is free and easy. Go to www.twitter.com, click one of the Sign Up buttons, and follow the onscreen instructions to create a new account.

Search for Users to Follow

Tweets from users you follow appear in your Home timeline. You can follow any Twitter user you want. (A user can be an individual, a company, or some other organization.)

Unlike Facebook, where friends and connections have to be mutually approved, you don't have to be approved to view another user's tweets. So if you want to follow Paul McCartney (@PaulMcCartney) or CNN (@CNN) or just your neighbor down the street, you can do so without having to ask permission.

The easiest way to find users to follow is to search for them.

@name

Users on Twitter are identified by a username preceded by an at sign (@). So, for example, my username is molehillgroup, which translates into my Twitter "handle" of @molehillgroup.

1. Click or tap within the search box and enter the name of the person or organization you want to follow.

2. Twitter displays a list of matching members. Click or tap a name to go to that user's profile page.

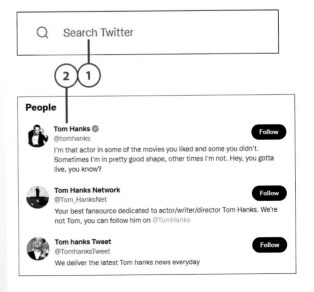

(3) Click or tap Follow to follow this user.

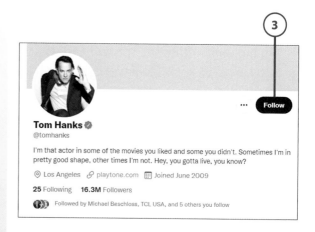

View Tweets

A tweet is a post to the Twitter service. Tweets from users you follow are displayed in Twitter's Home timeline.

(1) Click the Home icon to view the tweets in your timeline.

(2) Tweets are listed in reverse chronological order, with the newest tweets at the top. The name of the tweeter and how long ago the tweet was made are listed at the top of each tweet. Scroll down the page to view older tweets.

(3) Click or tap a user's name or @ name within the tweet to view the profile summary.

(4) Click or tap the Like (heart) icon to "like" a tweet.

(5) Click or tap the Reply icon to reply to a tweet.

(6) To view other tweets on a highlighted topic, click or tap the hashtag (#topic) within the tweet. (Not all tweets include hashtags.)

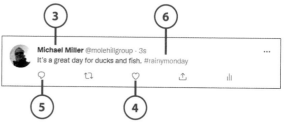

(7) To view a web page linked to within a tweet, click the embedded URL.

(8) Photos are embedded within tweets but at a limited size. To view the full picture, click it.

>>>Go Further

TWEET SHORT AND SWEET

Tweets are limited to 280 characters of text. Because of the 280-character limitation, tweets do not always conform to proper grammar, spelling, and sentence structure—and, in fact, seldom do.

It is common to abbreviate longer words, use familiar acronyms, substitute single letters and numbers for whole words, and refrain from all punctuation. For example, you might shorten the sentence "I'll see you on Friday" to read "CU Fri." You'll get used to it.

Post a Tweet

Posting a message to Twitter is called *tweeting*. The posts you make are called *tweets*.

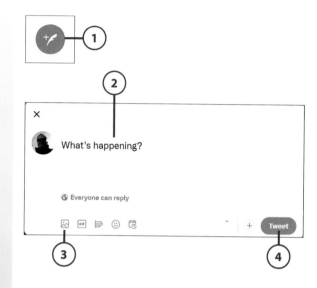

1. From anywhere on the Twitter website, click or tap the Tweet button. This displays the Compose New Tweet panel.

2. Type your message into the What's Happening? box. Remember that a tweet can be no more than 280 characters in length.

3. Click or tap the Picture icon to add a picture or video to your tweet.

4. Click or tap the Tweet button when you're done.

>>>Go Further
HASHTAGS

On Twitter, a *hashtag* is a word or phrase (with no spaces) in a tweet that is preceded by the hash or pound character, like this: **#hashtag**. Hashtags function much like keywords by helping other users find relevant tweets when searching for a particular topic. A hashtag within a tweet is clickable; clicking a hashtag displays a list of the most recent tweets that include that word.

To add a hashtag to a tweet, simply add the hash character (#) before a specific word. Most Twitter users include at least one hashtag in every tweet.

Retweet Another Tweet

Sometimes you'll see a tweet from some person or organization in your feed that you'd like to share with your friends. You do this by *retweeting* the original tweet. (The new tweet you create is called a *retweet*.)

1. From the Twitter home page, click or tap the Retweet icon for the tweet you want to share.

2. Click or tap Retweet to retweet without additional comment. *Or…*

3. Click or tap Quote Tweet to retweet with a new comment.

4. Enter your comments (optional) into the Add a Comment box.

5. Click or tap the Tweet button.

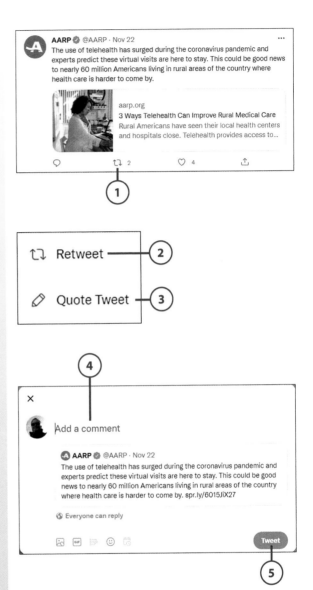

>>>*Go Further*

OTHER POPULAR SOCIAL MEDIA

Facebook, Pinterest, and Twitter are the three most popular social networks among older users, but they aren't the only social networks out there. Other social media networks cater to different demographic groups—and might be worth considering if you want to connect to younger friends or family members.

Some of the other popular social networks include the following:

- Instagram (www.instagram.com), which is designed for photo and video sharing

- LinkedIn (www.linkedin.com), which targets business professionals and is good for business networking and job hunting

- Snapchat (www.snapchat.com), which lets users share photos and short videos for short periods before the messages disappear

- TikTok (www.tiktok.com), which displays a seemingly random selection of short user videos that young people apparently find entertaining

These and most other social networks are free to use and have both web-based and mobile versions.

18

Storing, Editing, and Sharing Your Pictures

If you're like me, you take a lot of pictures with your smartphone and digital camera. You can use your Windows 11 computer to store, edit, and share those photos with friends and family—online, over the Internet.

Using Your Smartphone or Digital Camera with Your Windows PC

The first step in managing all your digital photos is to transfer those pictures from your digital camera or smartphone to your computer. There are a number of ways to do this.

Transfer Photos from the Cloud

If you're like me, you take most of your photos with your smartphone. It's certainly more convenient to whip out your phone to take a quick picture than it is to lug around a digital camera everywhere you go.

The photos you take with a smartphone can be stored in one of two places by default. Almost all phones store photos on the phone itself, although it's easy to run out of storage space if you take a lot of pictures. Most phones also offer the option of storing photos in the cloud, where they can be accessed from any device connected to the Internet—including your computer.

If you have an Apple iPhone or iPad, iCloud is the default cloud storage service. When your device is properly configured, all the photos you take are automatically transferred from your phone or tablet to the cloud. To view and download the photos you take, all you have to do is access the iCloud website at www.icloud.com with your Apple account.

If you have an Android phone or tablet, Google Photos is the cloud storage of choice. Google Photos works just like iCloud, automatically backing up all the photos you take with your phone to the cloud. You can then download photos to your computer by going to the Google Photos website at photos.google.com.

Transfer Photos Directly from a Smartphone or Tablet

It's also easy to transfer photos stored on your smartphone or tablet directly to your PC. All you need is the connection cable supplied with your device.

1. Connect one end of the supplied cable to your smartphone or tablet.

(2) Connect the other end of the cable to a USB port on your PC. If you receive an Autoplay notification, follow the instructions to view the photos on your phone. Otherwise, proceed to step 3.

(3) Click or tap File Explorer on the taskbar or Start menu to open File Explorer.

(4) Click or tap the This PC icon in the navigation pane.

(5) Click or tap the icon for your smartphone or tablet.

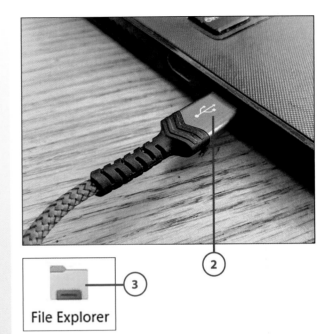

File Explorer

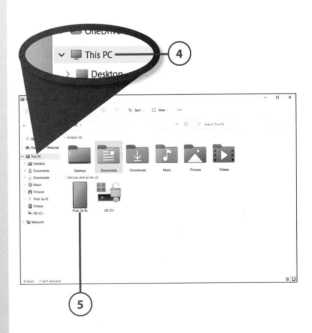

6. Navigate to the main folder on the device (typically labeled DCIM or Pictures) and then select the appropriate subfolder to see your photos.

7. Hold down the Ctrl key and click or tap each photo you want to transfer.

8. Click or tap the Copy button on the toolbar.

9. Click or tap the Pictures icon in the Navigation pane to open the Pictures folder.

10. Click or tap the Paste button on the toolbar. This copies all the selected photos to the Pictures folder on your computer.

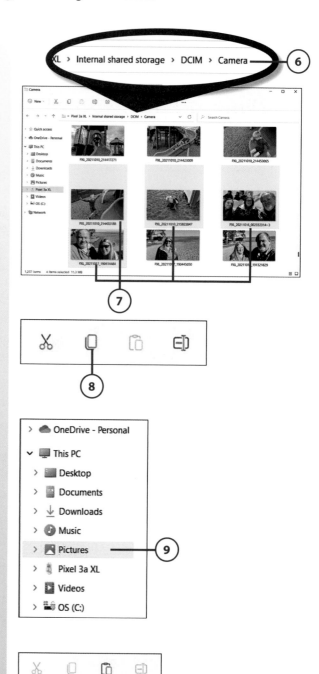

Download from the Your Phone App

If you have an Android smartphone, you can use the Your Phone app to link your two devices and view and download photos stored on your phone. Learn more in Chapter 16, "Using Your Windows PC with Your Android Phone."

Transfer Photos from a Memory Card

If you still use a digital camera to take photos, it's equally easy to transfer your pictures from your camera to your computer. The easiest way to do this is to use your camera's memory card.

Connecting Your Camera Directly

You can transfer photos by connecting your digital camera to your computer via USB. This is similar to connecting your smartphone or tablet to your PC; Windows should recognize when your camera is connected and automatically download the pictures in your camera while displaying a dialog box that notifies you of what it's doing.

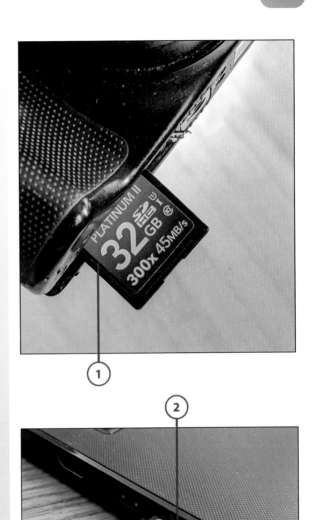

(1) Turn off your digital camera and remove the flash memory card.

(2) Insert the memory card from your digital camera into the memory card slot on your PC.

Copying Automatically

Windows might recognize that your memory card contains digital photos and start to download those photos automatically—no manual interaction necessary. Alternatively, you might get prompts from any other photo app you have installed to download your photos to that app.

(3) Click or tap File Explorer on the taskbar or Start menu to open File Explorer.

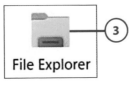

File Explorer

4 Click or tap This PC in the navigation pane.

5 Click or tap the icon for your memory card.

6 Navigate to the main folder on the memory card (typically labeled DCIM or Pictures) and then select the appropriate subfolder to see your photos.

7 Hold down the Ctrl key and click each photo you want to transfer.

8 Click or tap Copy on the toolbar.

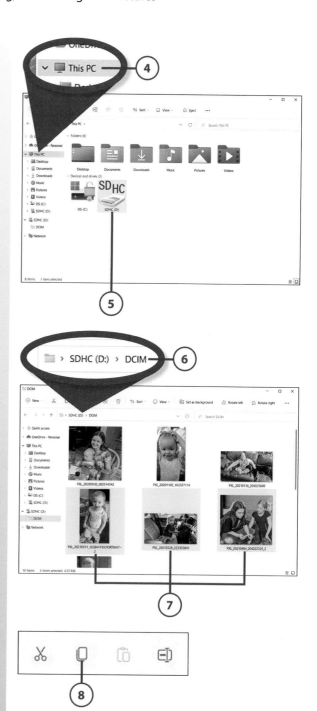

(9) Click or tap the Pictures icon in the Navigation pane to open the Pictures folder.

(10) Click or tap the Paste button on the toolbar. This copies all the selected photos to the Pictures folder on your computer.

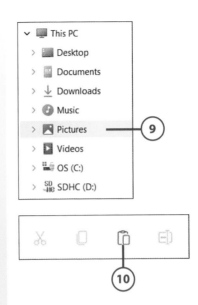

Delete Your Photos

Once you've copied photos from your smartphone or camera to your computer, you can delete them from your device. If you don't delete them, your phone or camera storage fills up rather quickly.

Viewing Photos on Your PC

Windows includes a built-in Photos app for viewing and editing photos stored on your PC.

View Your Photos

The Photos app is the hub for all your photo viewing and editing in Windows. It lets you navigate to and view all the photos stored on your PC. You launch the Photos app from the Start menu.

(1) Within the Photos app, the Collection view is selected by default and photos are grouped by date taken. To instead display pictures stored in specific folders on your computer's hard drive, select Folders view.

(2) Change the size of the photos displayed by clicking either View Large, View Medium, or View Small.

(3) Display a single photo within the Photos app by clicking it.

(4) Move to the next picture by clicking or tapping the right arrow on the screen or pressing the right-arrow key on your keyboard. To return to the previous picture, click or tap the left arrow on the screen or press the left-arrow key on your keyboard.

(5) Zoom into or out of the picture by clicking or tapping the Zoom In or Zoom Out buttons.

(6) Click or tap Delete (or press the Del key on your keyboard) to delete the current picture.

(7) Click or tap Rotate to rotate the picture 90 degrees clockwise.

(8) Click or tap See More (three-dot icon) and then select Slideshow to view a slideshow of the pictures in this folder, starting with the current picture.

(9) Click or tap the back arrow to return to the previous screen.

>>>*Go Further*

LOCK SCREEN AND BACKGROUND PICTURE

To use the current picture as the image on the Windows Lock screen, display the photo full screen, click See More (three-dot icon), click Set As, and then click Set As Lock Screen. To set this picture as your desktop background, click See More, click Set As, and then click Set As Background.

Create and View Photo Albums

The Photos app enables you to organize your photos in virtual photo albums. Let's look first at how to create and view a photo album.

1. From within the Photos app, click Albums to display the Albums view.

2. All of your previously created albums are displayed here. Click or tap an album to view the photos within.

3. Click or tap the New Album tile to create a new album.

4. Click or tap to select the photos you want to include.

5. Click or tap Create.

(6) Name this album by highlight-
ing the "Album" title and typing a
new title.

OneDrive

By default, your photo albums are stored
locally on your current PC. If you want your
albums to be available to other devices via
Microsoft's OneDrive cloud storage service,
open an album and click or tap the Save to
OneDrive button.

Touching Up Your Photos

Not all your pictures turn out perfect. Maybe you need to crop a picture to high-
light the important area. Maybe you need to brighten a dark picture or darken a
bright one. Or maybe you need to adjust the tint or color saturation.

Fortunately, the Windows Photos app lets you do this sort of basic photo editing. A better-looking photo is only a click or a tap away!

Enter Editing View

All of the Photo app's editing functions
are accessed via a special editing view.
Here's how you enter editing view.

(1) From within the Photos app, navigate to and display the photo you
want to edit.

(2) Click or tap the Edit Image button
to display your photo in editing
view, ready to edit.

Rotate a Photo

Is your picture sidewise? To turn a portrait into a landscape, or vice versa, use the Photos app's Rotate tool.

1. Enter editing view and click or tap Crop & Rotate.

2. Click or tap Rotate to rotate the picture 90 degrees clockwise. Continue clicking or tapping to further rotate the picture.

3. To rotate in less than 90-degree increments, click or tap and drag the Straightening control until the picture is in the desired position.

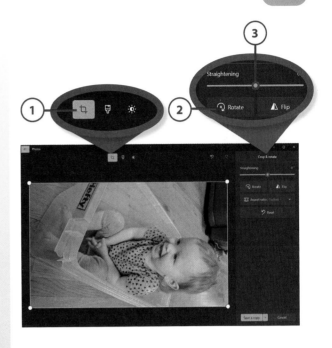

Crop a Photo

Sometimes you don't get close enough to the subject for the best effect. When you want to zoom in closer, use the Photos app's Crop control to crop out the edges you don't want.

1. Enter editing view and click or tap Crop & Rotate.

2. If you want to crop to a specific aspect ratio, click or tap Aspect Ratio and make a selection— Custom, Original, Square, 3:2, 4:3, 7:5, or 10:8.

3. Use your mouse or the touchpad to drag the corners of the white border until the picture appears as you like.

Apply a Filter

The Photos app includes several built-in filters you can apply to your pictures. Use filters to apply interesting effects quickly and easily to a photo.

1. Enter editing view and click or tap to display the Filters tab.

2. Click or tap the filter you want to apply from the Choose a Filter section.

3. Drag the slider to increase (right) or decrease (left) the intensity of the filter.

Remove Red Eye

Red eye is caused when a camera's flash causes the subject's eyes to appear a devilish red. The Photos app lets you remove the red eye effect by changing the red color to black in the edited photo.

1. Enter editing view and click or tap to select the Adjustments tab.

2. Scroll down and click or tap Red Eye. The cursor changes to display a translucent blue circle.

3. Click the eye(s) you want to fix to remove the red eye effect.

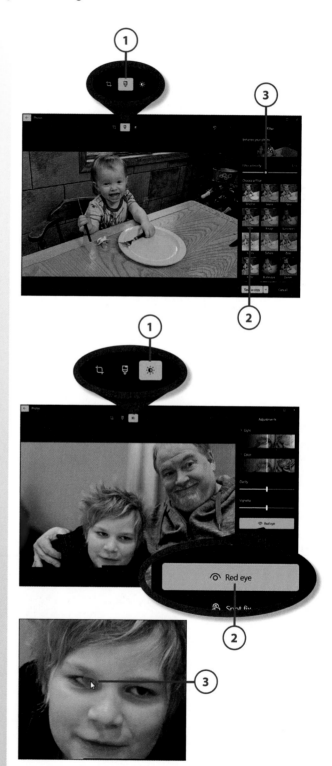

Retouch a Photo

Does someone in your photo have a blemish or a loose hair? Is there a rough or scratched area in the photo you want to get rid of? Or is there a bit of drool dripping down a cute baby's chin? Use the Photos app's Retouch control to smooth out or remove blemishes from your photos.

1. Enter editing view and click or tap to select the Adjustments tab.

2. Scroll down and click or tap Spot Fix. The cursor changes to include a translucent blue circle.

3. Click or tap the area you want to repair. The area is now repaired.

Adjust Brightness and Contrast

When a photo is too dark or too light, use the Photos app's Light controls. The Contrast control increases or decreases the difference between the photo's darkest and lightest areas. The Exposure control increases or decreases the picture's exposure to make the overall picture lighter or darker. Use the Highlights control to bring out or hide detail in too-bright highlights; use the Shadows control to do the same in too-dark shadows.

1. Enter editing view and click or tap to select the Adjustments tab.

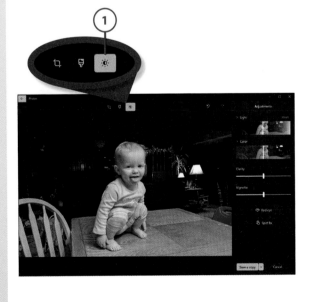

(2) In the Light section, click or tap and drag the white line to the left to make the picture darker or to the right to make the picture lighter.

(3) To display additional brightness and contrast controls, click or tap Light.

(4) Click or tap and drag the control for the item you want to adjust— Contrast, Exposure, Highlights, or Shadows.

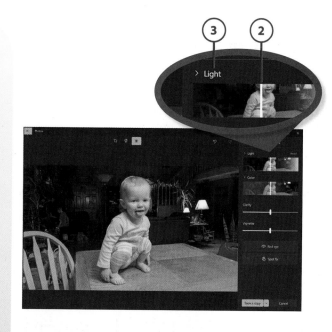

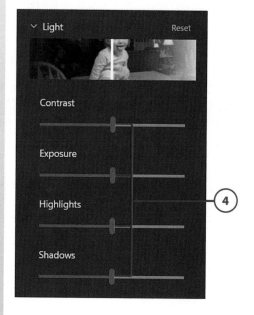

Adjust Color and Tint

The Photos app lets you adjust various color-related settings.

1. Enter editing view and click or tap to select the Adjustments tab.

2. In the Color section, click or tap and drag the white line to the left to decrease the color saturation for the picture or to the right to increase the color saturation.

3. To display additional color-related controls, click or tap Color.

4. Click or tap and drag the Tint control to change the tinting of the picture.

5. Click or tap and drag the Warmth control to the left to make the picture cooler (more blue) or to the right to make a warmer (more red) picture.

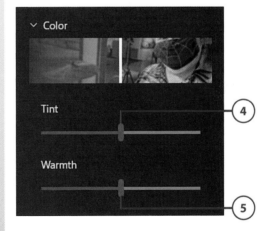

Apply Other Effects

You can also use the Photos app to change the picture's focus or apply a vignette effect.

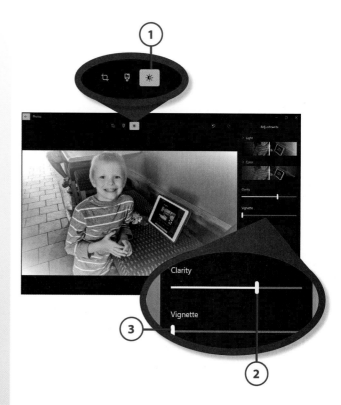

1. Enter editing view and click or tap to select the Adjustments tab.

2. Click or tap and drag the Clarity control to the left to make the picture more blurry or to the right to make it sharper.

3. Click or tap and drag the Vignette control to the left to apply a white vignette around the picture or to the right to apply a black vignette.

Save Your Work

You can opt to save your changes to the original picture or as a copy of that picture—which leaves the original unchanged.

1. From the editing view, click or tap Save a Copy to save your changes to a new file, leaving the original file unchanged. *Or...*

2. Click or tap the down arrow and select Save to save your changes to the current file.

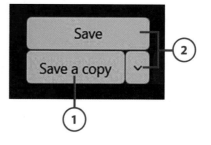

Cancel Changes

If, when you're editing a photo, you decide you don't want to keep the changes you've made, click or tap the Cancel button.

>>>*Go Further*
OTHER PHOTO-EDITING PROGRAMS

If you need to edit your photos beyond what you can do in the Photos app, you need to install and use a more full-featured photo-editing program. These programs let you do everything you can in the Photos app and a lot more—for a price.

Some of the more popular photo-editing programs include Adobe Photoshop Elements (www.adobe.com) and Corel PaintShop Pro (www.paintshoppro.com/en/products/paintshop-pro/). Both programs offer a variety of photo-editing tools, and sell for less than $100.

Sharing Your Pictures

It's fun to look at all the digital pictures you've stored on your PC, but it's even more fun to share those items with family and friends. Fortunately, the Internet makes it easy to share your favorite photos online, so everyone can ooh and aah over your cute children or grandchildren.

Sharing a Photo from the Photos App

If you're working from within the Photos app, Windows makes it easy to share a photo a number of different ways.

1. Open the photo you want to share and then click or tap the See More (three-dot) button.

2. Click or tap Share to open the Share panel.

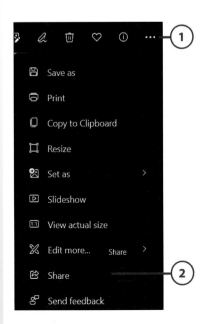

(3) Select the app you want to use to share and follow the normal procedure for that app to select a recipient, add a text message, and send the photo.

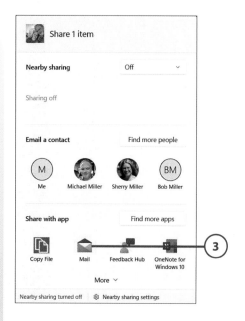

Attach a Photo in the Mail App

If you're working within the Windows Mail app, you can use that app to send one or more photos via email. You do this by attaching a picture file to an email message and then sending that item along with the message to your intended recipients. A friend or family member who opens your email can click to view the photo.

All email programs and services let you attach photo files to your messages. Here's how it's done in the Mail app:

(1) Launch the Mail app and then click or tap + New Mail to open a new email message.

(2) Enter the recipient and subject information as normal.

(3) Enter any accompanying text into the message area.

(4) Click or tap the Insert tab.

(5) Click or tap Files to open the Open window.

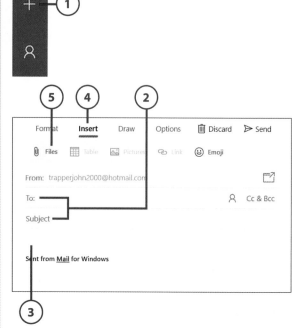

6. Navigate to and select the photo(s) you want to share.

7. Click or tap Open.

8. A thumbnail for the photo appears in your message.

9. Click or tap Send to send the message to its recipients.

Other Email Programs

Other email programs and services, such as Gmail and Yahoo! Mail, also let you attach photos in a similar fashion.

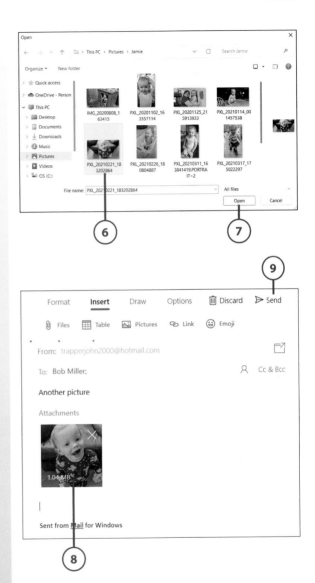

Watching Movies and TV Shows on Your PC

Want to rewatch last night's episode of *The Voice*? Or the entire season of *NCIS*? How about a classic episode of *The Dick Van Dyke Show* or *Dragnet*? Or maybe one of those prestige streaming series, such as *Only Murders in the Building*, *Stranger Things*, or *The Wheel of Time*? What about that latest "viral video" you've been hearing about?

You're probably used to watching TV shows on your TV. Thanks to streaming video technology and the Internet, however, you can watch all your favorite programs and movies on your computer, via your web browser. Assuming you have a broadband Internet connection, you can find tens of thousands of free and paid videos to watch through dozens of different streaming video services, including Amazon Prime, Disney+, Hulu, Netflix, and YouTube.

Watching Streaming Video Services

Streaming video is video programming that's transmitted either live or on-demand over the Internet. You can watch it on just about any type of device, from so-called smart TVs and streaming media players to smartphones, tablets, and your friendly neighborhood Windows 11 PC.

Some streaming video services are free, but most require a paid subscription (anywhere from $5 to $15 per month, depending on the service). Once you sign up, you can then watch all available videos from that service on any Internet-connected device—including your Windows 11 PC.

Amazon Prime Video

You probably think of Amazon as a big online retailer, which it is. Amazon is also a big player in streaming video, with its Amazon Prime Video service.

Like other paid streaming video services, Amazon Prime Video offers a mix of original and existing programming, both TV series and movies. Some of Amazon's most popular original series include *The Boys*, *Good Omens*, *Hunters*, *The Marvelous Mrs. Maisel*, and *The Wheel of Time*. In addition to its Prime Video offerings, Amazon also offers a variety of on-demand programming for sale or rental.

You watch Amazon Prime Video in your web browser on Amazon's website (www.amazon.com). A Prime Video subscription costs $8.99 USD per month, or it's free if you already have an Amazon Prime membership.

Amazon Prime

Amazon Prime Video is both separate from and a part of Amazon Prime, the service that gives you free shipping on many Amazon orders. It's separate in that you can subscribe separately, but if you have a full Amazon Prime membership ($12.99 USD per month or $119 USD per year), you not only get free shipping on your Amazon purchases, but you also get Prime Video for free. Depending on how much shopping you do at Amazon, that might be the best deal.

Apple TV+

Apple TV+ is Apple's streaming video service. It offers a selection of original series—including *For All Mankind, Foundation, The Morning Show, Ted Lasso,* and *Schmigadoon!*—as well as some theatrical movies.

You watch Apple TV+ in your web browser on Apple's website (www.apple.com/apple-tv-plus/). It costs $4.99 USD per month.

Free Trials

Most paid video streaming services offer some sort of free trial period. If you sign up for a free trial and don't like the service, remember to cancel your subscription before you get hit with the first month's billing.

Discovery+

Discovery+ is the streaming home for reality and do-it-yourself programming from A&E, Animal Planet, Discovery, DIY Network, Food Network, HGTV, History Channel, Lifetime, Magnolia Network, OWN, Science Channel, TLC, and the Travel Channel. You watch Discovery+ in your web browser on the Discovery+ website (www.discoveryplus.com). A subscription runs $4.99 USD per month.

Discovery+ Plus HBO Max Equals…

As I write this, Discovery Networks is in the process of acquiring AT&T's media properties, all of which are included in the current HBO Max streaming service. It is likely that Discovery will integrate some or all of HBO Max into Discovery+ for some sort of super-duper streaming service, so look for that to happen later in 2022.

Disney+

Disney+ offers streaming content from properties owned by the Walt Disney media conglomerate—specifically movies and television series from Disney, Marvel, National Geographic, Pixar, and 20th Century Fox. This includes all the Marvel superhero movies, programming from the Disney Channel and Disney Jr., classic and current Disney movies, and all the *Star Wars* movies and series.

That's a lot of programming, which makes it a favorite service for many, especially if you have younger viewers, comic book fans, or *Star Wars* followers in your household. Original Disney+ programming includes *The Beatles: Get Back, Doogie Kamealoha M.D., The Falcon and the Winter Soldier, Hawkeye, The Mandalorian, Under the Helmet: The Legacy of Boba Fett,* and *WandaVision.*

You watch Disney+ in your web browser on the Disney+ website (www. disneyplus.com). A subscription costs just $7.99 USD per month or $79.99 USD per year.

HBO Max

HBO Max offers content from all the properties owned by parent company AT&T—Cartoon Network, CNN, The CW, DC Comics, HBO, New Line Cinema, Turner Classic Movies (TCM), and the Warner Bros. film studio. Despite the HBO name, HBO Max is designed to be a full-service streaming service, much like Netflix, with a mix of existing and original programming.

Some of the more popular HBO Max originals include *The Flight Attendant, Game of Thrones, Hacks, Mare of Easttown, Succession,* and *Titans*. HBO Max is also home of the *Sex and the City* revival, *And Just Like That....*

You watch HBO Max in your web browser on the HBO Max website (www.hbomax.com). A monthly subscription runs $9.99 USD with commercials or $14.99 USD without commercials, although you may get a discount if you also subscribe to HBO through your cable provider.

Hulu

Hulu is a streaming service that offers a mix of movies, television shows, and original programming. It is particularly known for its selection of current episodes from the major TV networks; it's a great place for catching up on any recent shows you've missed.

Original programming available on Hulu includes *American Horror Story*, *Dopesick, The Great, The Handmaid's Tale, Letterkenny, Only Murders in the Building,* and *Reservation Dogs*. Hulu is also a good place to find a variety of vintage television programs.

Hulu offers two subscription plans, with and (sort of) without commercials. The basic Hulu plan runs $6.99 USD per month and inserts commercials into the programs you watch. If you want to minimize the number of commercials you see, sign up for the $12.99 USD No Ads plan—but know that you'll still see commercials on some programs, although fewer of them.

You watch Hulu in your web browser on the Hulu website (www.hulu.com).

Netflix

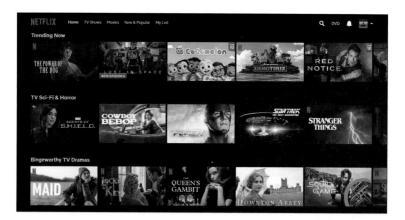

When it comes to watching movies and TV shows online, the most popular streaming service today is Netflix (www.netflix.com). Netflix offers a broad variety of television shows, movies, and original programming—perhaps the biggest selection of any competing service.

DVD Rental

Netflix also offers a separate DVD-by-mail rental service, which has a separate subscription fee. That's not what I'm talking about here, however.

Netflix's original programming includes *Cobra Kai, Fuller House, Grace and Frankie, The Kominsky Method, Ozark, Squid Game, Stranger Things, Sweet Tooth, Tiger King, The Umbrella Academy,* and *The Witcher.*

Netflix offers three different subscription plans. The Basic plan costs $9.99 USD per month and offers only standard definition viewing on a single screen or device. The more popular Standard plan costs $15.49 USD per month and offers HD viewing on two simultaneous screens or devices. The Premium plan runs $19.99 USD per month and offers up to 4K definition on up to four screens or devices at the same time.

You watch Netflix in your web browser on the Netflix website (www.netflix.com).

SD, HD, and 4K

The resolution of a TV picture measures how detailed the picture is. Older TV sets and computer monitors can only display standard resolution (SD). Newer TVs and computer screens can display high definition (HD). Some very new TV sets (but few if any computer screens) can display 4K resolution. If you're watching Netflix on your computer, the Standard plan with HD resolution is all you need.

Paramount+

Paramount+, formerly known as CBS All Access, offers content from the CBS television network (including CBS news and sports), as well as BET, Comedy Central, MTV, Nickelodeon, Paramount Pictures, and the Smithsonian Channel. (All channels are owned by the CBSViacom conglomerate.)

Paramount+ is also the home for a variety of original programming, including *The Good Fight, Mayor of Kingstown, The Stand, Yellowstone,* and all the new *Star Trek* shows (including *Star Trek: Discovery, Star Trek: Picard,* and *Star Trek: Strange New Worlds*).

There are two monthly subscription plans. The Essential plan runs $4.99 USD per month or $49.99 USD per year. If you want to excise advertisements and livestream your local CBS station, go with the Premium plan for $9.99 USD per month or $99.99 USD per year.

As with all of these streaming services, you watch Paramount+ in your web browser on the Paramount+ website (www.paramountplus.com).

Peacock

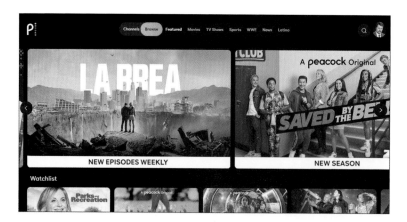

Peacock is NBCUniveral's streaming video service, with programming from the NBC, Syfy, and USA networks, as well as movies from Universal Pictures. Peacock also offers a variety of original programming, including shows like *Battlestar Galactica, Girls5Eva, MacGruber, One of Us Is Lying, Rutherford Falls,* and the *Saved by the Bell* revival.

You watch Peacock in your web browser on the Peacock website (www.peacocktv.com). As for subscriptions, Peacock offers a free version with commercials and limited programming; a Premium version with all the service's programming and all the ads, for $4.99 USD per month; and a Premium Plus version with no ads for $9.99 USD per month.

Comcast/Xfinity Subscribers

Because Peacock and NBCUniversal are owned by Comcast, if you get your cable or Internet service from Comcast/Xfinity, you may be able to get a discount on the Peacock service.

>>>*Go Further*

PURCHASING AND RENTING ON-DEMAND VIDEOS

Although most people today get their entertainment from "all you can eat" streaming video services, such as Hulu and Netflix, you can still purchase and rent movies and TV shows either for downloading to your PC or for on-demand streaming. This type of on-demand viewing is most typically used for watching movies just out of the theaters and not yet available on the streaming video services.

The most popular sources for on-demand videos are Amazon (www.amazon.com), Google Play (play.google.com), Redbox (www.redbox.com), and Vudu (www.vudu.com). Prices vary from a few dollars apiece for individual TV episodes to $20 or so for a new-release movie.

Watch Other Paid Streaming Video Services

There are a number of more targeted paid streaming video services you can watch on your Windows 11 computer. All of these services offer monthly or yearly subscriptions and are viewable from any web browser. These services include:

- Acorn TV (www.acorn.tv, $5.99 USD per month), offering a variety of television programming from Britain, Ireland, Australia, and New Zealand

- BET+ (www.bet.plus, $9.99 USD per month), movies and TV shows from Black creators for Black audiences

- BritBox (www.britbox.com, $6.99 USD per month), home to British TV programs from BBC and ITV

- BroadwayHD (www.broadwayhd.com, $11.99 USD per month), with a variety of Broadway plays and musicals

- The Criterion Channel (www.criterionchannel.com, $10.99 USD per month), the perfect service for movie buffs, with classic and foreign films from the Criterion Collection

- ESPN+ (plus.espn.com, $6.99 USD per month), with thousands of live sporting events from the various ESPN channels

Watch Free Streaming Video Services

Although the most popular streaming services tend to be subscription-based, there are also several free streaming video services you can watch on your Windows 11 computer. These services offer a variety of (typically older) movies and TV shows, complete with commercials.

The most popular free streaming video services today include the following:

- Crackle (www.crackle.com)
- IMDb TV (www.imdb.com)
- Plex (www.plex.tv)
- Pluto TV (www.pluto.tv)
- Popcornflix (www.popcornflix.com)
- The Roku Channel (therokuchannel.roku.com)
- Tubi (www.tubitv.com)
- Xumo (www.xumo.tv)

>>>*Go Further*

NETWORK TV PROGRAMMING

Most major broadcast and cable TV networks offer their shows for viewing (for free) on their own websites. All you have to do is fire up your web browser and start watching. The most popular of these network TV sites include ABC (abc.go.com), CBS (www.cbs.com), Comedy Central (www.comedycentral.com), CW (www.cwtv.com), Fox (www.fox.com), NBC (www.nbc.com), Nick (www.nick.com), TNT (www.tntdrama.com), and USA Network (www.usanetwork.com).

Watching Live TV on Your PC

There's a ton of great programming available via the major streaming video services, but there's still a lot of good stuff on good ol' broadcast television, too. Is there a way to watch your local channels, broadcast networks, and cable channels on your computer? Yes, there is—by subscribing to a live TV streaming service.

These services, such as Hulu with Live TV, Sling TV, and YouTube TV, offer a selection of live programming from a variety of sources that you can watch over the Internet in your computer's web browser. Most offer live streaming of a wide variety of cable programming; many also offer live streaming of your local TV channels.

Navigating these live TV streaming services is a lot like using a cable or satellite channel guide. You typically see a list of channels down the left side of the guide and upcoming times (in half-hour increments) along the top. You scroll down the guide to select a channel and scroll right to see what's coming up in the near future.

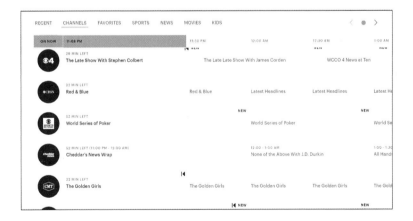

The channel guide for Hulu + Live TV.

These live TV streaming services aren't free—although they cost a lot less than a typical cable and satellite plan. Prices often run around $60 USD per month or more, although you can customize your plans to some degree to view more or fewer channels.

In addition, most of these live TV streaming services offer cloud DVR functionality. That is, you can record your favorite shows just as you would with a cable or satellite box, and the shows are stored on the service's cloud servers somewhere on the Internet. You can watch your recorded programs with the click of a button; many of these services also offer on-demand program for your viewing pleasure.

fuboTV

fuboTV's original focus was streaming live worldwide sporting events, and it's still one of the best plans for sports fans with more sports channels than its competitors. It offers more than just sports, however, with the typical assortment of cable channels and local channels in most areas.

As to pricing, the fubo Starter plan offers 116 channels and a 250-hour cloud DVR for $64.99 USD per month. If you want more recording time, go with the Pro package at $69.99 USD per month to get 1,000 hours of cloud DVR recording. For more channels (and the extra hours of recording), the Elite plan offers 162 channels at $79.99 USD per month.

Learn more at www.fubo.tv.

Hulu + Live TV

Hulu + Live TV is an extension of the traditional Hulu streaming service. This service offers more than 75 live cable and local channels and 50 hours of cloud DVR storage. Pricing starts at $69.99 USD per month for a bundle that includes Live TV, regular Hulu, Disney+, and ESPN+ (all owned by Disney).

Learn more at www.hulu.com/live-tv.

Philo

Philo differs from the other live TV streaming services in that it doesn't offer any local channels. You get more than 60 cable channels and unlimited cloud DVR storage (for 30 days), but you only pay $25 USD per month. That's considerably lower than competing services, which makes it an attractive service to pick up any channels you don't get otherwise.

Learn more at www.philo.com.

Sling TV

Sling TV is the granddaddy of live TV streaming services. The company sells two plans: Sling Orange has 32 channels and costs $35 USD per month, whereas Sling Blue has 43 channels and also costs $35 USD per month. (You can subscribe to both Orange and Blue for $50 USD per month.) Both plans come with 10 hours of cloud DVR storage. The company also offers a variety of extra channel add-ons.

Unlike other live TV streaming services, you can't watch Sling TV in your browser. Instead, you have to download and install the Sling TV app from the Microsoft Store. The app is free and easy to use. Learn more at www.sling.com.

YouTube TV

YouTube TV (not to be confused with the traditional YouTube video-sharing service, discussed later in this chapter) is fast becoming the live TV streaming service of choice. It offers a wide variety of programming (85+ channels) and cloud DVR recording with unlimited storage. YouTube TV also offers the greatest number of local channels in the most locations.

You pay $64.99 USD per month for a subscription that includes just about everything, although you can pay more to add some premium channels. Learn more at tv.youtube.com.

>>>Go Further

WATCHING STREAMING VIDEO ON YOUR LIVING ROOM TV

Watching movies and TV shows on your PC is fine if you're on the go, but it's not the same as watching programming on the big flat-screen TV you have in your living room or bedroom. Some newer "smart" TVs have built-in Internet connectivity, so you can watch Amazon Prime Video, Disney+, Hulu, Netflix, and other services directly from the TV itself.

For those "non-smart" TVs that don't have built-in Internet connectivity, you need to connect some sort of streaming media box or "stick" to your TV to watch streaming video. The three most popular devices are Amazon Fire TV, Google Chromecast, and Roku, all of which connect directly to your TV's HDMI input.

Another option is to connect your computer to the TV in your bedroom or living room. You can connect via an HDMI cable or, in some instances, wirelessly by "casting" the signal from your computer to your TV. Learn more about connecting your PC to your TV in Chapter 5, "Connecting Printers and Other Peripherals."

Viewing and Sharing Videos on YouTube

There's another place to watch videos online, and it's a big one. YouTube, unlike the streaming video services just discussed, doesn't specialize in commercial TV shows and movies. Instead, YouTube is a video-sharing community; users can upload their own videos and watch videos uploaded by other members.

YouTube is where you find all those homemade videos of cute cats and laughing babies that everybody's watching, as well as tons of "how-to" videos, video blogs, videogame tutorials, and more. And when you find a video you like, you can share it with your friends and family—which is what helps a video go "viral."

View a Video

You access YouTube from any web browser. Unlike many of the streaming video services we've discussed, you use YouTube for free—no subscription necessary.

Movies on YouTube

In addition to its user-uploaded videos, YouTube offers a variety of commercial movies. Some movies are free; others can be rented on a 48-hour pass for as low as $1.99 USD.

---

1 From within your web browser, go to the YouTube site at www.youtube.com. Browse through the suggested videos. *Or…*

2 Search for a particular video by entering what you're looking for into the Search box and then pressing Enter or clicking the Search (magnifying glass) button.

3 Click or tap the video you want to watch.

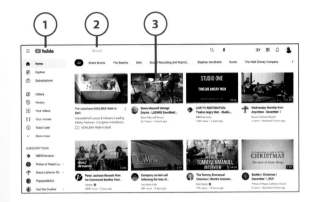

4 The video begins playing automatically when the video page displays.

5 Click or tap the Pause button to pause playback; click or tap the button again (now a Play button) to resume playback.

6 Click or tap the Fullscreen button to view the video on your entire computer screen.

7 Click or tap the thumbs-up button to "like" the video, or the thumbs-down button to "dislike" it.

>>>Go Further

SHARING VIDEOS

Find a video you think a friend would like? YouTube makes it easy to share any video with others.

Click or tap the Share button under the video player to display the Share panel. You can then opt to email a link to the video, "like" the video on Facebook, or tweet a link to the video on Twitter.

Upload Your Own Video

If you take movies with your camcorder or smartphone, you can transfer those movies to your computer and then upload them to YouTube. This is a great way to share your home videos with friends and family online. (You have to be signed into YouTube before you start uploading, of course.)

Uploading from a Smartphone

If you shoot video with your smartphone or tablet, you can probably upload to YouTube directly from your device or the YouTube mobile app. Check your device or app to see what options are available.

1. Click the Create button at the top of any YouTube page.

2. Select Upload Video.

3. Click or tap Select Files to display the Open dialog box.

4. Navigate to and select the video file you want to upload.

5. Click or tap the Open button.

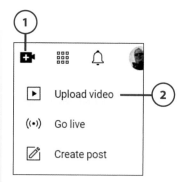

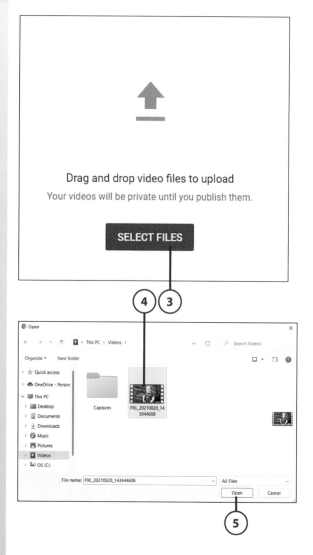

6 As the video is uploaded, YouTube displays the video information page. Follow the onscreen instructions to enter information about the video, including title and description.

In this chapter, you find out how to listen to music on your PC, from Pandora, Spotify, and other streaming music services, as well as music you purchase and download from Amazon and Google Play Music. You also discover how to find and listen to entertaining and informative podcasts on your PC.

20

Listening to Music and Podcasts on Your PC

Many folks like to listen to music and podcasts on their computers. You can stream music and podcasts over the Internet to your computer or purchase and download tunes from online music stores and play them back on your PC. Whichever way you choose to listen, your computer—and Windows—will do the job.

Listening to Streaming Music

People our age have been conditioned to purchase the music we like, whether on vinyl, cassette tape, compact disc, or via digital download. But there's an entire world of music on the Internet that you don't have to purchase. It's called *streaming music*, and it gives you pretty much all you can listen to for a low monthly subscription price—or even for free. There's nothing to download; the music is streamed to your computer in real time, over the Internet.

The two largest streaming music services are Pandora and Spotify. We'll look at each of these.

>>>*Go Further*

ON-DEMAND VERSUS PERSONALIZED SERVICES

There are two primary types of delivery services for streaming audio over the Internet. The first model, typified by Pandora, is like traditional radio in that you can't dial up specific tunes; you have to listen to whatever the service beams out but in the form of personalized playlists or virtual radio stations. The second model, typified by Spotify, lets you specify which songs you want to listen to; these are *on-demand services*.

Know, however, that even services that focus on one type of streaming often offer the alternative approach. For example, Pandora's free service offers a radio-like approach that doesn't let you dial up individual songs, but Pandora also offers a paid Premium service that offers personalized on-demand listening.

Listen to Pandora

Pandora is much like traditional AM or FM radio in that you listen to the songs Pandora selects for you, along with accompanying commercials. It's a little more personalized than traditional radio, however, because you can create personalized stations. All you have to do is choose a song or artist; Pandora then creates a station with other songs like the one you picked. You access Pandora from the company's website (www.pandora.com) or from the Pandora Windows app, which is available for free download from the Microsoft Store.

Free Versus Paid

Pandora's basic membership is free, but ad-supported. (You have to suffer through commercials every few songs.) To get rid of the commercials, pay for the $4.99 USD per month Pandora Plus subscription. Or sign up for the $9.99 USD per month Pandora Premium, which lets you personalize your music with on-demand selections.

1. To use Pandora on the Web, go to www.pandora.com and sign up or log in.

2. Click or tap the Browse tab to browse through featured play-lists and other recommendations. *Or…*

3. Click or tap My Collection to view stations you've previously created.

4. Click or tap a station or playlist to start playing it.

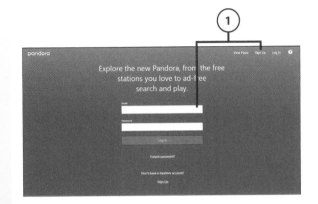

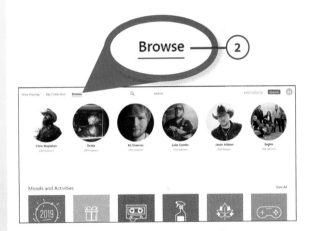

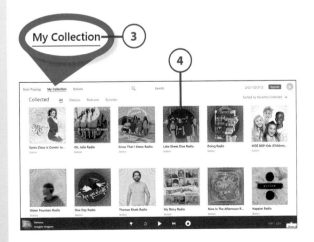

5 Click or tap the Now Playing tab to view the currently playing item. Information about the current track and artist is displayed.

6 Click or tap Pause to pause playback. Click or tap Play to resume playback.

7 "Like" the current song by clicking or tapping the thumbs-up icon. Pandora will play more songs like this one.

8 If you don't like the current song, click or tap the thumbs-down icon. Pandora skips to the next song, won't play the current one again, and will play fewer songs like it.

9 Skip to the next song without disliking it by clicking or tapping the Next Track button.

10 To create a new station, enter the name of an artist, song, or genre into the Search box and then press Enter.

11 Pandora lists artists, albums, songs, stations, playlists, and podcasts that match your query. Click or tap the one you want, and Pandora starts playback and adds it to your list of favorites.

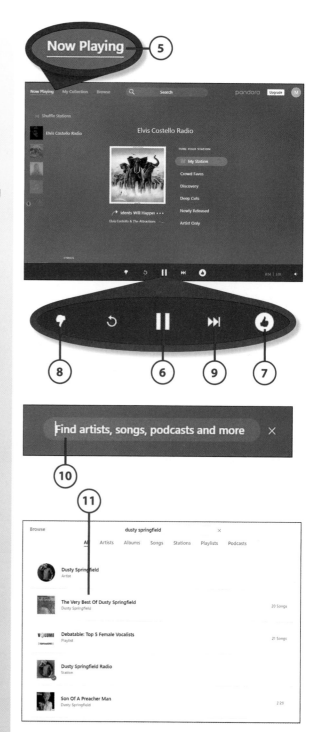

>>>Go Further

LOCAL RADIO STATIONS ONLINE

If you'd rather just listen to your local AM or FM radio station—or to a radio station located in another city—you can do so over the Internet. There are three major radio services online: Audacy (www.audacy.com), iHeartRadio (www.iheart.com), and TuneIn (www.tunein.com). All offer free access to local radio stations around the world. (You have to search the services to determine which offers the specific station you're looking for; most stations are exclusive to one service.)

Listen to Spotify

The other big streaming music service today is Spotify. Unlike Pandora, Spotify's paid service lets you choose specific tracks to listen to.

Spotify offers a web-based version you can access via your web browser or a standalone app that offers enhanced functionality. Access both at www.spotify.com.

Free Versus Paid

Spotify's basic membership is free, but you're subjected to commercials every few songs, and the service sometimes inserts its choices into your playlists. If you want to get rid of the commercials (and get true on-demand playback), you need to pay for a $9.99 USD per month subscription.

1. Go to www.spotify.com and either sign up or log into your account.

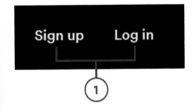

(2) Click or tap to play any of the suggestions on the home page. *Or…*

(3) Click or tap Your Library to view playlists you've previously created. *Or…*

(4) Click Search to browse through music by genre or mood.

(5) Enter the name of a song, album, or artist into the Search box and then press Enter.

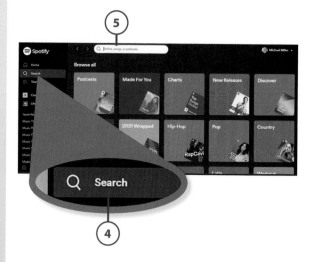

6. Click the green Play button to play all the songs in the playlist or album or by that artist.

7. Double-click a song title to play that particular track.

8. Use the playback controls at the bottom to pause, rewind, or fast-forward playback.

>>>Go Further
OTHER STREAMING MUSIC SERVICES

Pandora and Spotify aren't the only streaming music services on the Internet. You can check out and listen to any of these popular streaming music services using your web browser:

- Amazon Music Prime (www.amazon.com/music/prime), free to Amazon Prime members
- Amazon Music Unlimited (music.amazon.com), $9.99 USD per month or $7.99 USD per month if you're an Amazon Prime member
- Apple Music (www.apple.com/apple-music/), $9.99 USD per month
- LiveXLive (www.livexlive.com), with three plans: Basic (free), Plus ($3.99 USD per month), and Premium ($9.99 USD per month)
- Napster (us.napster.com), $9.99 USD per month
- TIDAL (www.tidal.com), with three plans: Free (free, with just "good" sound quality), HiFi ($9.99 USD per month, with high fidelity sound), and HiFi Plus ($19.99 USD per month, with even better lossless high-fidelity sound)
- YouTube Music (music.youtube.com), $9.99 USD per month

In addition, SiriusXM lets you listen to all of its satellite-based stations (and additional online-only stations) over the Internet. Their online service costs $10.99 USD per month, or it's free if you have a SiriusXM subscription for your car.

Purchasing Digital Music Online

Prior to streaming music services, the only way to get music online was to purchase and download individual tracks or complete albums from an online music store. The two biggest online music stores today are Amazon Digital Music and Apple's iTunes Store. Because Apple's store requires you to download and install the iTunes software to make a purchase, this section focuses on the Amazon store, which you can access from any web browser.

Purchase Music from the Amazon Digital Music Store

The Amazon Digital Music Store is a major source of downloadable digital music in MP3 format. You can play the music you purchase from Amazon's online store in any music playback app, including the Windows 11 Media Player app, which is covered later in this chapter.

Amazon offers tens of millions of tracks for purchase. Prices run from 69 cents to $1.29 USD per track, with complete albums also available.

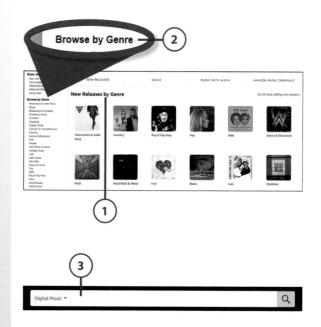

1. Point your web browser to mp3. amazon.com. The main page displays new and recommended releases.

2. Browse by genre by scrolling down to the Browse by Genre section in the left column; then click or tap the genre you want.

3. Search for a specific song, album, or artist by entering your query into the Search box at the top of the page and clicking or tapping the magnifying glass button or pressing Enter.

4 Click to select the artist or album you want.

5 Purchase an entire album by clicking the Buy MP3 Album button. (Alternatively, click the Add to MP3 Cart button to purchase more than one item at this time.)

6 Purchase an individual track by clicking the down arrow for that track and then selecting Buy Song. (Alternatively, select Add Song to MP3 Cart to purchase more than one item at this time.)

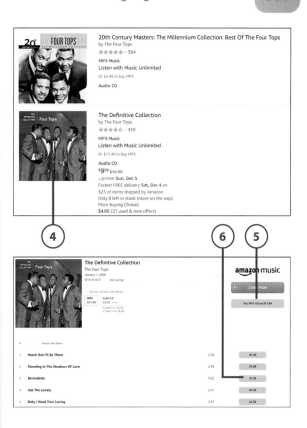

Listen to Digital Music with the Windows 11 Media Player App

How do you listen to the music you've downloaded from the Amazon store? There are several music player apps available, but the easiest one to use is the one that's included with Windows 11—the Media Player app.

You launch the Media Player app from the Windows Start menu.

1 Select Music Library in the navigation pane on the left.

2 Select Songs to view the individual tracks in your music library.

3 Select Albums to view the albums in your music library.

4 Select Artists to view your music sorted by performing artist.

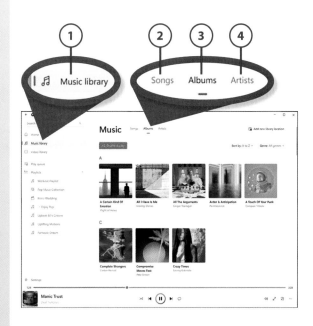

(5) Double-click or double-tap the artist, album, or track you want to play.

(6) Playback controls are displayed at the bottom of the Media Player window. Click or tap Pause to pause playback or Play to resume playback.

(7) Click or tap Shuffle to play the tracks by an artist or album in random order.

Listening to Podcasts Online

A podcast is like a traditional radio program, but it streams over the Internet. There are all types of podcasts available—music podcasts, true crime podcasts, political podcasts, you name it.

Most podcasts are free and available from most major streaming music services, such as Spotify and TuneIn Radio. You can also find podcasts on dedicated podcast websites, such as Audible and Google Podcasts.

Find Podcasts with Google Podcasts

I'm using Google Podcasts as an example because it's easy to use and available from any web browser. Create a free account or sign in with your Google account at podcasts.google.com.

1 On the Google Podcasts home screen, scroll down to view top podcasts by type. *Or…*

2 Search for a specific podcast by typing its name into the Search for Podcasts box.

3 Click or tap the podcast in which you're interested.

4 Click or tap the Play button to listen to a specific episode.

5 Click or tap the Subscribe button to subscribe to and get notification of future episodes of this podcast.

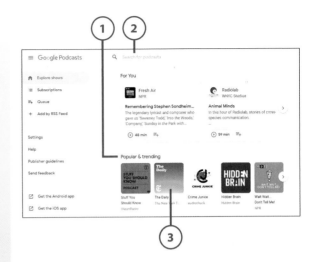

Listen to a Podcast

Once you find a podcast you like, listening to that podcast on your computer is as easy as clicking your mouse.

(1) Click or tap the Play button for the podcast to which you want to listen. This displays the playback controls at the bottom of the screen.

(2) Click or tap the Pause button to pause playback. The Pause button now changes to a Play button; click or tap this button to resume playback.

(3) Click or tap the 10 button to skip backward 10 seconds.

(4) Click or tap the 30 button to skip forward 30 seconds.

(5) Drag the playback slider to move to another point in the podcast.

(6) To change the playback speed (so you can listen faster or slower), click or tap the 1.0X button and select a different speed.

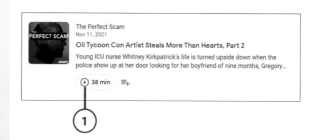

The Perfect Scam
Nov 11, 2021

Oil Tycoon Con Artist Steals More Than Hearts, Part 2

Young ICU nurse Whitney Kirkpatrick's life is turned upside down when the police show up at her door looking for her boyfriend of nine months, Gregory...

38 min

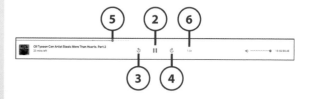

>>>Go Further

POPULAR PODCASTS

To get you started with the wide world of podcasts, here are a dozen of the most popular podcasts in which you may be interested, all available from Google Podcasts:

- *99% Invisible*, which looks at design in the world around us.
- *Better Health While Aging*, which discusses common health problems that affect those of us over age 60.
- *Criminal*, which examines real-world criminal cases from the nineteenth century to today.
- *The Daily*, a daily 20-minute news podcast produced by the *New York Times*.
- *Freakonomics Radio*, the top-rated podcast among older listeners, all about economics in your daily life.
- *Good Job, Brain!*, a popular interactive trivia quiz show.
- *Lux Radio Theater*, presenting hundreds of hour-long classic radio dramas originally aired between 1934 to 1955.
- *The Not Old—Better Show*, aimed at listeners aged 50+ by those 50+.
- *The Perfect Scam*, a weekly podcast from AARP that focuses on understanding and avoiding both online and real-world cons and scams.
- *TED Radio Hour*, a fascinating journey through technology, entertainment, and design, produced by NPR.
- *This American Life*, entertaining true stories presented by host Ira Glass.
- *You Must Remember This*, which uncovers the fascinating behind-the-scenes stories of classic Hollywood.

These are just a few of the tens of thousands of podcasts available—browse through Google Podcasts to find many more!

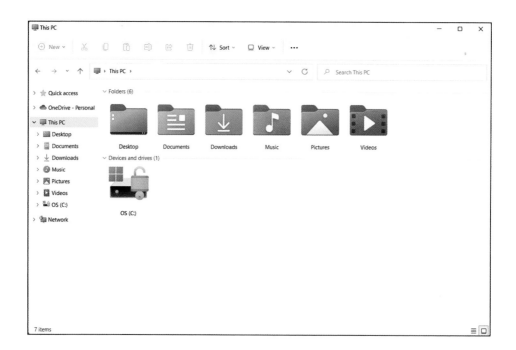

In this chapter, you see how to use File Explorer to manage the files and folders on your PC's hard drive.

→ Using File Explorer
→ Working with Folders
→ Managing Files
→ Working with Microsoft OneDrive

21

Using Files and Folders

All the data for documents and programs on your computer is stored in electronic files. A file can be a word processing document, a music track, a digital photograph—just about anything, really.

The files on your computer are organized into a series of folders and subfolders. It's just like the way you organize paper files in a series of file folders in a filing cabinet—only it's all done electronically.

Using File Explorer

You might, from time to time, need to work with the files on your computer. You might want to copy files from an external USB memory drive, for example, or move a file from one folder to another. You might even want to delete unused files to free up space on your hard drive.

When you need to manage the files on your Windows 11 computer, you use an app called File Explorer. This app lets you view and manage all the files and folders on your PC—and on connected devices.

File Explorer in Windows 11

File Explorer got a bit of an overhaul in Windows 11. In addition to new file icons, the old ribbon is replaced by a new toolbar with the most common file commands. This provides a simplified interface with the most-used commands available at the click of a mouse.

There are a few ways to open File Explorer:

- Click or tap the File Explorer icon on the taskbar.

- Click or tap the Start button to open the Start menu and then click or tap the File Explorer icon.

- Right-click or right-tap the Start button to open the Options menu; then click or tap File Explorer.

- Press Windows+E on your computer keyboard.

Navigate Folders and Libraries

All the files on your computer are organized into folders. Some folders have subfolders—that is, folders within folders. There are even sub-subfolders, and sub-sub-subfolders. It's a matter of nesting folders within folders, in a kind of hierarchy. Naturally, you use File Explorer to navigate the various folders and subfolders on your PC's hard disk.

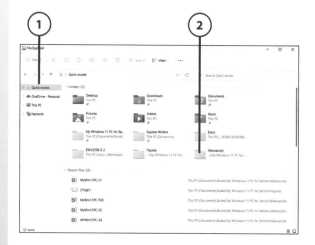

1 In File Explorer's default view, Quick Access is selected, and your most-used folders and documents are displayed. Double-click or double-tap any item to view the contents.

2 A given folder may contain multiple folders and subfolders. Double-click or double-tap any item to view its contents.

3 To move back to the disk or folder previously selected, click or tap the Back button on the toolbar.

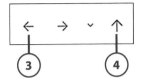

4 To move up the hierarchy of folders and subfolders to the next highest item, click or tap the up-arrow button on the toolbar.

>>>Go Further
BREADCRUMBS

File Explorer includes an Address box at the top of the window, which displays your current location in terms of folders and subfolders. This list of folders and subfolders presents a "breadcrumb" approach to navigation; it's like leaving a series of breadcrumbs behind as you delve deeper into the hierarchy of subfolders.

> « Documents › Books › My Windows 11 PC for Seniors ›

You can view additional folders within the hierarchy by clicking or tapping the separator arrow next to the folder icon in the Address box. This displays a pull-down menu of the recently visited and most popular items.

Use the Navigation Pane

Another way to navigate your files and folders is to use the navigation pane on the left side of the File Explorer window. This pane displays both favorite links and hierarchical folder trees for your computer, libraries, and networks.

1 Click or tap the right arrow next to the This PC icon in the navigation pane to expand and display the folders within.

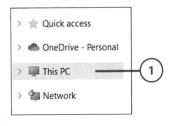

2. Click or tap the right arrow next to any folder to expand and display all the subfolders it contains.

3. The right arrow changes to a down arrow. Click or tap this to hide the expanded subfolders.

4. Click or tap an icon in the navigation pane to open the contents of the selected item.

Change the Folder View

You can choose to view the contents of a folder in a variety of ways. File Explorer lets you display files as Small Icons, Medium Icons, Large Icons, or Extra Large Icons. You also have the option of displaying files as Tiles, Details, or a List. There's even a Content view that displays information about the file beside it.

1. From within File Explorer, click or tap View in the toolbar to open the drop-down menu. Select one of the options described in steps 2 through 6.

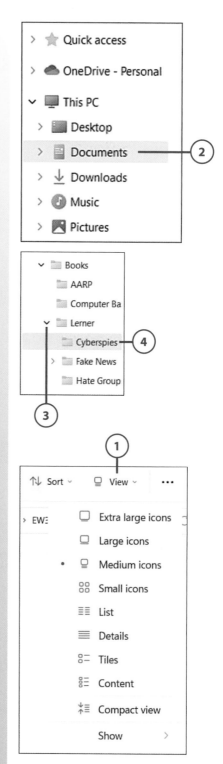

2 Content displays files with content descriptions.

3 Details displays columns of details about each file.

4 List displays files in a simple list.

2

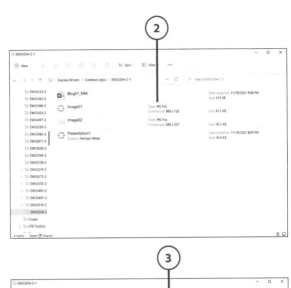

3

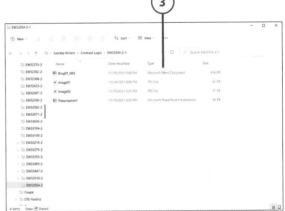

4

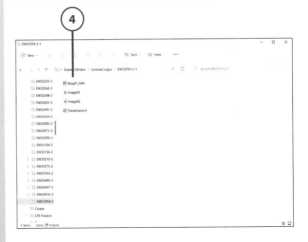

5 Tiles displays files as small tiles.

6 Small Icons, Medium Icons, Large Icons, or Extra Large Icons displays files as icons of the corresponding size.

7 Click Show to access options to display the Details pane (details about the selected file), the Preview pane (contents of the selected file), Item Check Boxes, File Name Extensions, and Hidden Items (typically system files you shouldn't be accessing).

Hide the Navigation Pane

You can also opt to click the Show option and hide the navigation pane, although doing so would make File Explorer considerably more difficult to use.

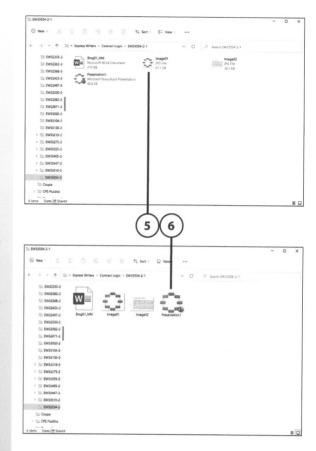

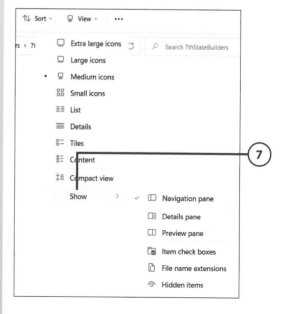

Sort Files and Folders

When viewing files in File Explorer, you can sort your files and folders in a number of ways. To view your files in alphabetic order, choose to sort by Name. To see all similar types of files grouped together, choose to sort by Type. To sort your files by the date and time they were last edited, select Date Modified.

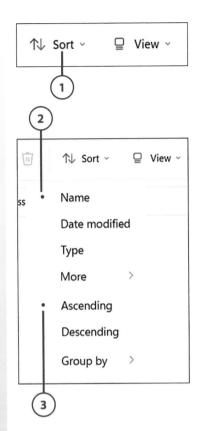

1. From within File Explorer, click or tap Sort on the toolbar to display the drop-down menu.

2. Choose to sort by Name, Date Modified, Type, or More (Size, Date Created, Authors, Categories, Tags, or Title).

3. By default, Windows sorts items in ascending order. To change the sort order, click or tap Descending.

Different Sorting Options

Different types of files have different sorting options. For example, if you're viewing music files, you can sort by Album, Artists, Bit Rate, Composers, Genre, and the like.

Working with Folders

Windows stores files in virtual folders. You can create new folders to hold new files or rename existing folders if you like.

Create a New Folder

The more files you create, the harder it is to organize and find things on your hard disk. When the number of files you have becomes unmanageable, you need to create more folders—and sub-folders—to better manage those files.

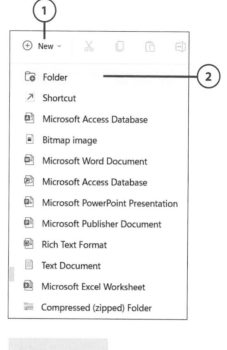

(1) From within File Explorer, navigate to the drive or folder where you want to place the new folder; then click or tap New on the toolbar to display the drop-down menu.

(2) Click or tap Folder.

(3) A new, empty folder appears with the filename New Folder highlighted. Type a name for your folder and then press Enter.

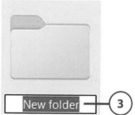

It's Not All Good

Wait!

When creating a new folder, do not press Enter or click the folder until you've entered a new name for it. Clicking the folder or pressing Enter locks in the current name as New Folder. You would then have to rename the folder (as described next) to change that name.

Rename a Folder or File

When you create a new folder, it helps to give it a name that describes its contents. Sometimes, however, you might need to change a folder's name. You may also want or need to change the name of an individual file. Fortunately, Windows makes renaming an item relatively easy.

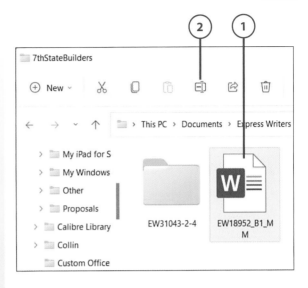

1. Click or tap the file or folder you want to rename.

2. Click or tap Rename on the toolbar; this highlights the filename.

3. Type a new name for your folder (which overwrites the current name) and then press Enter.

Keyboard Shortcut

You can also rename a folder or file by selecting the item and pressing F2 on your computer keyboard. This highlights the name and readies it for editing.

>>>Go Further

WHY ORGANIZE YOUR FOLDERS?

You might never have an occasion to open File Explorer and work with your files and folders. But there's some value in doing so, especially when it comes to organizing your files.

Perhaps the best example of this is when you have a large number of digital photos—which, if you're a grandparent, like me, you probably do. Instead of lumping hundreds or thousands of photos into a single Photos folder, you can instead create different subfolders for different types of photos. For example, you might want to create folders named Vacation Photos, Family Photos, and Holiday Photos, or something similar.

Personally, I like organizing my photos by year and month. Within my main Photos folder, I have subfolders for 2018, 2019, 2020, 2021, and the like. Then, within each year's folder, I have subfolders for each month—January, February, March, and such. This way, I can quickly click through the folders to find photos taken in a particular month.

You can organize your photos and other files similarly or use whatever type of organization suits you best. The point is to make all of your files easier to find, however you choose to do so.

Managing Files

Tens of thousands of files are stored on a typical personal computer. From time to time, you might need to manage them in various ways. You can copy a file to create a duplicate in another location, or you can move a file from one location to another. You can even delete files from your hard drive, if you like. And you do all this with File Explorer.

Copy a File

Copying a file places a duplicate of the original file into a new location. There are many ways to copy a file in Windows; the easiest is to use the Copy command on the toolbar.

1. From within File Explorer, navigate to and click or tap the item you want to copy.

2. Click or tap Copy on the toolbar.

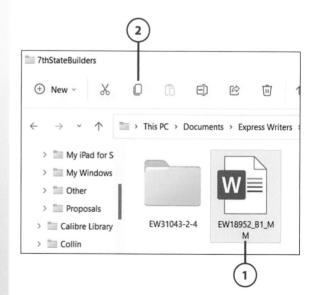

(3) Navigate to and open the folder where you want to copy the file.

(4) Click or tap Paste on the toolbar. A copy of the file is placed in this location.

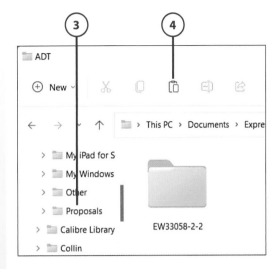

Move a File

Moving a file or folder is different from copying it. Moving cuts the item from its previous location and pastes it into a new location. Copying leaves the original item where it was and creates a copy of the item elsewhere.

(1) From within File Explorer, navigate to and click the item you want to move.

(2) Click or tap Cut on the toolbar.

(3) Navigate to and open the folder where you want to move the file.

(4) Click Paste on the toolbar. The file is pasted into this new location.

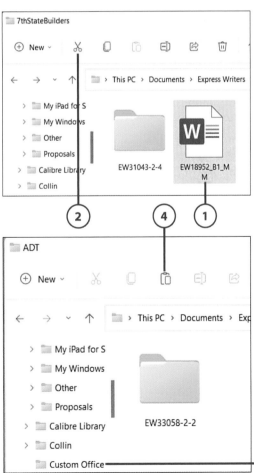

Delete a File or Folder

Keeping too many files eats up a lot of hard disk space on your computer—which can be a bad thing. Because you don't want to waste disk space, you should periodically delete those files (and folders) you no longer need. When you delete a file or folder, you send it to the Windows Recycle Bin, which is kind of a trash can for deleted files.

1. From within File Explorer, navigate to and click or tap the item you want to delete.

2. Click or tap Delete on the toolbar.

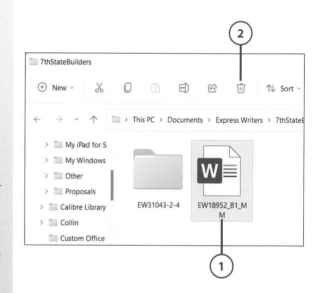

Other Ways to Delete

You can also delete a file or folder by dragging it from the File Explorer window onto the Recycle Bin icon on the desktop, or you can select it and press the Delete key on your computer keyboard.

Restore a Deleted File

Have you ever accidentally deleted the wrong file? If so, you're in luck. Windows stores the files you delete in the Recycle Bin, which is actually a special folder on your hard disk. For a short period of time, you can "undelete" files from the Recycle Bin to put them back to their original locations—and save yourself from making a bad mistake.

1. On the Windows desktop, double-click or double-tap the Recycle Bin icon to open the Recycle Bin folder.

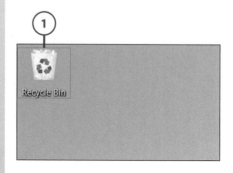

(**2**) Click or tap the file you want to restore.

(**3**) Click or tap the See More (three-dot) button on the toolbar.

(**4**) Select Restore the Selected Items.

Empty the Recycle Bin

By default, the deleted files in the Recycle Bin can occupy 4GB plus 5% of your hard disk space. When you've deleted enough files to exceed this limit, the oldest files in the Recycle Bin are automatically and permanently deleted from your hard disk. You can also manually empty the Recycle Bin and thus free up some hard disk space.

(**1**) From the Windows desktop, double-click or double-tap the Recycle Bin icon to open the Recycle Bin folder.

(**2**) Click or tap Empty Recycle Bin on the toolbar.

(**3**) When prompted, click or tap Yes to permanently delete these files.

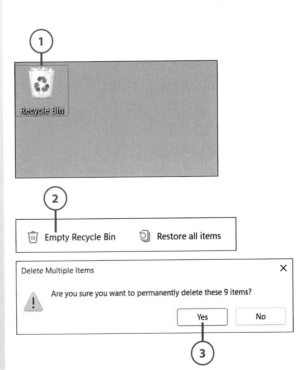

Working with Microsoft OneDrive

Microsoft offers online storage for all your documents and data via its OneDrive service. When you store your files on OneDrive, you can access them via any computer or mobile device connected to the Internet.

Cloud Storage

Online file storage, such as that offered by OneDrive, Apple's iCloud, and Google Drive, is called *cloud storage*. The advantage of cloud storage is that you can access files from any computer or other device at any location—work, home, or away. You're not limited to using a given file on one particular computer.

Manage OneDrive Files on the Web

Because OneDrive stores your files on the Web, you can manage all your OneDrive files with your web browser, from any Internet-connected computer. You can access OneDrive online from your web browser at onedrive. live.com or from the OneDrive icon in the notification area of the toolbar.

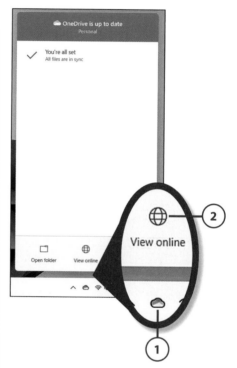

1. Click or tap the OneDrive icon in the notification area of the toolbar.

2. Click or tap View Online. This takes you to the OneDrive website in your web browser.

3. Your OneDrive files are stored in folders. Click or tap a folder to view its contents.

4 Click or tap a file to view it or, in the case of an Office document, open it in its host application.

5 To copy, move, rename, or delete a file, select the file and then choose the desired option from the toolbar.

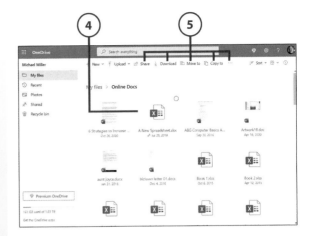

Storage Plans

Microsoft gives you 5GB of storage in your free OneDrive account, which is more than enough to store most users' documents, digital photos, and the like. If you need more storage, you can purchase 100GB of storage for $1.99 USD per month. (If you subscribe to the Microsoft 365 Personal plan, you get 1TB of storage for free.)

Manage OneDrive Files from File Explorer

You can also manage your online OneDrive files from the OneDrive folder in File Explorer.

1 From within File Explorer, click or tap to open the OneDrive - Personal folder to see all your OneDrive folders.

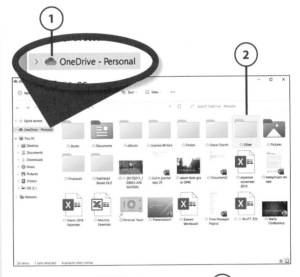

2 Click or tap a folder to view its contents.

3 Click or tap a file to view it or, in the case of an Office document, open it in its host application.

4 To cut, copy, rename, or delete a file, select the file and then select the action you want to perform from the File Explorer toolbar.

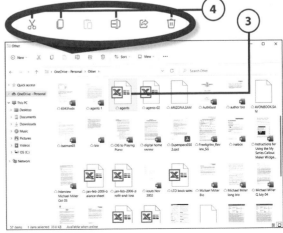

Upload a File to OneDrive

You can upload any file on your hard drive to OneDrive for storage online.

(1) On the OneDrive website, navigate to and open the folder where you want to store the file. (If you don't select a folder, the file will be uploaded to the main OneDrive directory.)

(2) Click or tap Upload on the toolbar.

(3) Click or tap Files to display the Open dialog box.

(4) Navigate to and select the file(s) you want to upload.

(5) Click or tap the Open button.

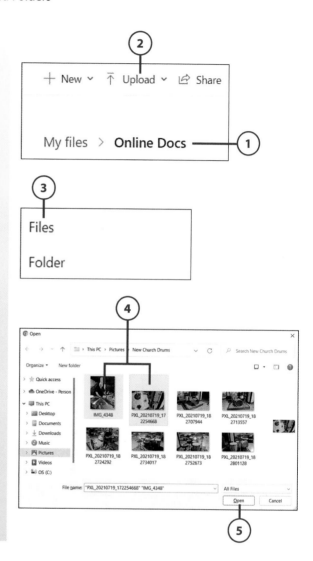

>>>Go Further

UPLOADING AND DOWNLOADING FROM FILE EXPLORER

You can also upload and download files to and from OneDrive from within File Explorer. To upload a file, copy that file from any other location to any folder within the OneDrive folder. To download a file, copy that file from the OneDrive folder to another location on your computer.

Download a File to Your PC

You can download files stored on OneDrive to your computer.

1. On the OneDrive website, select the file(s) you want to download.

2. Click or tap Download. When prompted to save the file, do so. (Unless you specify otherwise, files downloaded from OneDrive are saved into the Download folder on your computer.)

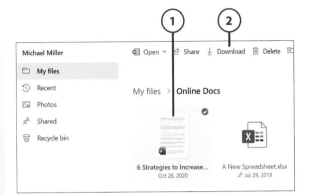

Task Manager

File Options View

Processes Performance App history Startup Users Details Services

Name	Status	2% CPU	37% Memory	0% Disk	0% Network	1% GPU	G
Apps (7)							
> 🌐 Google Chrome (7)		0%	225.4 MB	0 MB/s	0 Mbps	0%	
> 📧 Microsoft Outlook		0%	162.4 MB	0 MB/s	0 Mbps	0%	
> 📄 Microsoft Word		0%	80.2 MB	0 MB/s	0 Mbps	0%	
> 📄 Microsoft Word		0%	75.3 MB	0 MB/s	0 Mbps	0%	
> 📊 Task Manager		0.3%	26.5 MB	0 MB/s	0 Mbps	0%	
> 🗂 Windows Explorer		0.3%	85.9 MB	0.1 MB/s	0 Mbps	0.1%	
> 📱 Your Phone (3)	⏸	0%	36.8 MB	0 MB/s	0 Mbps	0%	
Background processes (87)							
> ☐ Adobe Acrobat Update Service ...		0%	0.3 MB	0 MB/s	0 Mbps	0%	
> ▣ Adobe Genuine Software Integri...		0%	1.7 MB	0 MB/s	0 Mbps	0%	
> ▣ Adobe Genuine Software Servic...		0%	0.7 MB	0 MB/s	0 Mbps	0%	
▣ AggregatorHost		0%	0.7 MB	0 MB/s	0 Mbps	0%	

∧ Fewer details   End task

In this chapter, you find out how to deal with common computer problems—and prevent those problems from happening.

→ Performing Necessary Maintenance
→ Fixing Simple Problems
→ Troubleshooting Other PC Problems

22

Dealing with Common Problems

Have you ever had your computer freeze on you? Or refuse to start? Or just start acting weird? Maybe you've had problems printing a document or opening a given program or finding a particular file. Or maybe you just can't figure out how to do a specific something.

Computer problems happen. When issues do occur, you want to get things fixed and running again as fast and as painlessly as possible. That's what this chapter is all about—dealing with those relatively common computer problems you might encounter.

Performing Necessary Maintenance

Before I get into dealing with fixing computer problems, let me explain how to prevent those problems. That's right—a little preventive maintenance can stave off a lot of future problems. Take care of your PC on a regular basis, and it will take care of you.

To ease the task of protecting and maintaining your system, Windows 11 includes several utilities to help you keep your computer running smoothly. You should use these tools as part of your regular maintenance routine—or if you experience specific problems with your computer system.

Automatically Clean Up Files with Storage Sense

Even with today's very large hard drive and SSD storage, you can still end up with too many useless files taking up too much storage space—especially if you're obsessed with taking vacation pictures or photos of your very cute grandkids. Fortunately, the Storage Sense utility can automatically delete temporary system and app files when your storage space runs low.

Storage Sense, when activated, runs in the background and springs into action when free disk space is running low—or, if you prefer, on a daily, weekly, or monthly schedule. It also deletes older files in your Recycle Bin and Downloads folder and, if you like, content that is duplicated in the cloud with OneDrive.

Settings

1. Open the Start menu and click or tap Settings to open the Settings app.

2. Click or tap to select System in the left column.

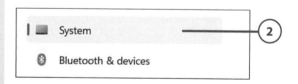

3. Click or tap Storage.

4. Scroll down and click or tap Storage Sense.

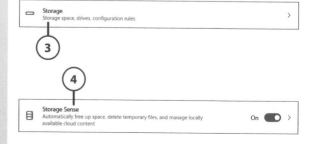

5 Make sure that the Cleanup of Temporary Files option is checked.

6 Click or tap "on" the Automatic User Content Cleanup switch.

7 Pull down the Run Storage Sense control and select when you want Storage Sense to run: During Low Disk Space (default), Every Day, Every Week, or Every Month.

8 Pull down the Delete Files in My Recycle Bin control and select a timeframe: Never, 1 Day, 14 Days, 30 Days (default), or 60 Days.

9 Pull down the Delete Files in My Downloads Folder control and select a timeframe: Never (default), 1 Day, 14 Days, 30 Days, or 60 Days. (This only deletes files that haven't changed in that time period.)

10 Pull down the OneDrive - Personal control and select a timeframe: Never (default), 1 Day, 14 Days, 30 Days, or 60 Days.

11 To run Storage Sense now (and delete those files marked for deletion), click or tap the Run Storage Sense Now button.

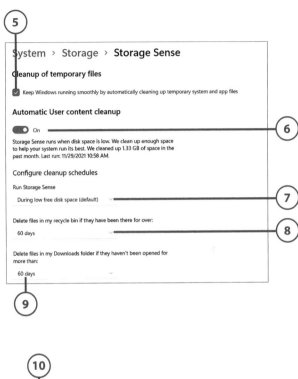

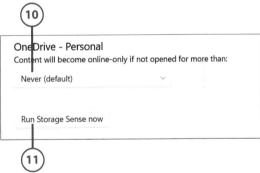

>>>Go Further
WHEN TO DELETE?

When configuring Storage Sense, how long should you keep files before you let Storage Sense delete them? For most users, 30 days is a good timeframe for all options.

When you're looking at files you've deleted and are stored in the Recycle Bin, you want to keep these files for a short time in case they were accidentally deleted and you need to restore them. Keeping a file in the Recycle Bin for more than 30 days is probably unnecessary.

In the case of files in your Downloads folder, 30 days is a good option, as well. If you haven't used a file you've downloaded within 30 days, you're probably never going to use it.

In the case of files stored both on your computer and on OneDrive, the choice is a little trickier. Some users want all their files available on their computer at all times, even when they're not connected to the Internet, so selecting Never is the best option for them. If you are more comfortable storing your files in the cloud, deleting local files after 30 days of nonuse (while still storing them online on OneDrive) is the right choice.

Manually Delete Unnecessary Files

Storage Sense runs automatically according to the options you configured. You can also opt to manually delete unnecessary files in your Recycle Bin and Downloads folder.

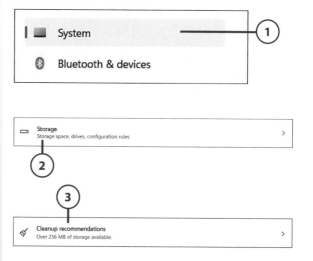

1. From within the Settings app, click or tap System in the left column.

2. Click or tap Storage.

3. Scroll down and click or tap Cleanup Recommendations.

4 Check the Downloads option to delete all files in your Downloads folder.

5 Check the Recycle Bin option to empty all deleted files from the Recycle Bin.

6 Click or tap the Clean Up button to remove the selected files.

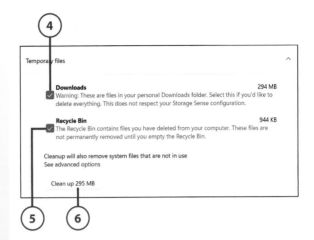

Large or Unused Files

You can also opt to delete especially large files or files that you haven't used in a long time. On the Cleanup Recommendations page in the Settings app, click or tap to expand the Large or Unused Files section. Check to select those files you want to delete and then click or tap the Clean Up button.

Delete Unused Programs

Another way to free up valuable hard disk space is to delete those programs you never use.

1 Within the Settings app, click or tap Apps in the left column.

2 Click or tap Apps & Features.

3 Find the program you want to delete and click or tap the More (three-dot) button.

4 Click or tap Uninstall.

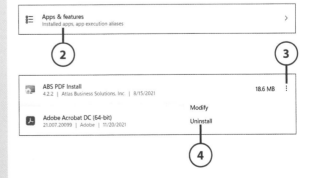

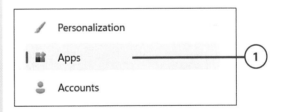

>>>*Go Further*

BACKING UP IMPORTANT FILES

The data stored on your computer's hard disk is valuable, and perhaps irreplaceable. We're talking about your personal photos, home movies, favorite music, spreadsheets, and word processing documents, and maybe even a tax return or two.

That's why you want to keep a backup copy of all these valuable files. The easiest way to store backup copies is on an external hard disk drive. These drives provide lots of storage space for a relatively low cost, and they connect to your PC via USB. There's no excuse not to do it!

When you're shopping for an external drive, get one at least as big as your PC's internal drive so you can copy your entire hard drive or SSD to the external drive. Then, if your system ever crashes, you can restore your backed-up files from the external drive to your computer's hard drive. External drives with 1TB or 2TB of storage are quite common today, typically selling for well under $100 USD—some under $50 USD.

You can find both external hard drives and external SSDs. External hard drives are a little lower cost than external SSDs, but SSDs are much faster when it comes to accessing files. They're also a lot more reliable over the long term.

Most external drives come with some sort of backup software installed, or you can use a third-party backup program. The backup process can be automated so that it occurs once a day or once a week and only backs up those files that are new or changed since your last backup.

Other users prefer to back up their data over the Internet, using an online backup service. This type of service copies your important files from your computer to the service's servers over the Internet. This way, if your local data is lost or damaged, you can restore the files from the online backup service's servers.

Several popular online backup services are designed for home users, including the following:

- Backblaze (www.backblaze.com)

- Carbonite (www.carbonite.com)

- IDrive (www.idrive.com)

The benefit of using an online backup service is that the backup copy of your library is stored offsite, so you're protected in case of any local physical catastrophe, such as fire or flood. Most online backup services also work in the background, so they're constantly backing up new and changed files in real time. Expect to pay $50 USD or more per year, per computer, for one of these services.

You can also use Microsoft's OneDrive or Google Drive for backup, although you'll probably need to invest in a pricier subscription to get enough storage space to do the job. However, these services aren't optimized for backup, which means there's more manual work involved in backing up all your files. Still, they're an affordable option.

Fixing Simple Problems

Computers aren't perfect—even new ones. It's always possible that at some point in time, something will go wrong with your PC. It might refuse to start; it might freeze up; it might crash and go dead. Then what do you do?

When something goes wrong with your computer, there's no need to panic (even though that's what you'll probably feel like doing). Most PC problems have easy-to-find causes and simple solutions. The key thing is to keep your wits about you and attack the situation calmly and logically.

You Can't Connect to the Internet

This problem is likely caused by a bad connection to your Wi-Fi network or hotspot. Fix the Wi-Fi problem and you can get back online lickity-split.

1. Try turning off and then turning back on your PC's wireless functionality. You might be able to do this from a button or switch on your computer, or you can do it within Windows. Click the Connections icon on the taskbar to display the Quick Settings pane; then click "off" the Wi-Fi control. Wait a few moments and then turn the Wi-Fi option back "on" and reconnect to your network.

2. It's possible that your computer is too far away from the wireless signal. Move your computer nearer to the closest Wi-Fi router or hotspot.

3. If you're using a public Wi-Fi hotspot, you might need to log on to the hotspot to access the Internet. Open your web browser and try to access any web page; if you're greeted with a log-in page for the hotspot, enter the appropriate information to get connected.

4. If nothing else works, it's possible that the hotspot to which you're trying to connect has Internet issues. Report your problem to whomever is in charge at the moment.

5. If you're on your home network, it's possible that your Wi-Fi router or cable modem (or combination gateway device) is the problem. Try turning off the router and modem or gateway device for five minutes or so, and then turning them back on.

6. It's also possible that your home Internet service provider (ISP) is having issues. If the problem persists, call your ISP and report the problem.

You Can't Go to a Specific Web Page

If you have a good connection to the Internet and can open some web pages, trouble opening a specific web page is probably isolated to that particular website.

1. The site might be having temporary connection issues. Refresh the web page to try loading it again.

2. You might have typed the wrong address for this particular site. Try entering the address again.

3. You might have the wrong address for a specific page on the website. Try shortening the address to include only the main URL—that is, go directly to the site's home page, not to an individual page on the site. For example, instead of going to www.informit.com/articles/article.aspx?p=2131141, just go to the main page at www.informit.com and navigate from there.

4. If you continue to have issues with this website, it's probably a problem with the site itself. That is, it's nothing you're doing wrong. Wait a few moments and try again to see if the problem is fixed.

You Can't Print

What do you do when you try to print a document on your printer and nothing happens? This problem could have several causes.

1. Click the Print button or command to open the Printer page or dialog box and then make sure the correct printer is selected.

2. Make sure the printer is turned on. (You'd be surprised…)

3. Check the printer to make sure it has plenty of paper and isn't jammed. (And if it is jammed, follow the manufacturer's instructions to unjam it.)

4. If your printer is connected via USB, check the cable between your computer and the printer. Make sure both ends are firmly connected. Lots of printer problems are the result of loose cables.

5. If you have a wireless printer, try uninstalling the printer from Windows and then reinstalling it. If that doesn't work, try to connect the printer via USB instead.

Your Computer Is Slow

Many computers will start to slow down over time. There are many reasons for this, from an overly full hard disk or SSD to an unwanted malware infection.

1. Close any open programs that don't need to be open at the moment.

2. Delete unnecessary files to free up hard disk space, as discussed in the "Manually Delete Unnecessary Files" section earlier in this chapter.

3. Install and run a reputable antimalware utility to find and remove any computer viruses or malware unknowingly installed on your system. (Learn more about antimalware utilities in Chapter 13, "Protecting Yourself Online.")

4. Ask a knowledgeable friend or professional computer technician to check your computer's startup programs; these are programs that load automatically when Windows starts up and run in the background, using valuable computer memory. Have your friend or technician remove those unnecessary startup programs.

Task Manager

You can view and manage your startup programs from the Task Manager utility. To open the Task Manager, press Ctrl+Alt+Del and then select Task Manager. Select the Startup tab to view those programs that launch during startup, and disable those you don't want to launch.

A Specific Program Freezes

Sometimes Windows works fine, but an individual software program stops working. Fortunately, Windows presents an exceptionally safe environment; when an individual application crashes or freezes or otherwise quits working, it seldom messes up your entire system. You can then use the Task Manager utility to close any frozen program without affecting other Windows programs.

1. When an application freezes, press Ctrl+Alt+Del.

2. Click the Task Manager option to launch the Task Manager utility.

3. Click the Processes tab. (You may need to click or tap More Details to see this tab.)

4. Go to the Apps section and click the program that's frozen.

5. Click the End Task button.

Your Entire Computer Freezes

If you're like many users, the worst thing that can happen is that your computer totally freezes, and you can't do anything—including shut it off. Well, there is a way to shut down a frozen computer and then restart your system.

1. Hold down the Windows key on your keyboard and simultaneously press your PC's power button. If that doesn't work, press and hold the PC's power button for several seconds, until the PC shuts down.

2. Wait a few moments and then turn your computer back on. It should restart normally. If not, you might need to consult a computer technician or repair service.

Troubleshooting Other PC Problems

No matter what kind of computer-related problem you're experiencing, you can take the following six basic steps to track down the cause of the problem. Work through these steps calmly and deliberately, and you're likely to find what's causing the current problem—and then be in a good position to fix it yourself:

1. Don't panic! Just because there's something wrong with your PC is no reason to get frustrated or angry. That's because it's likely that there's nothing seriously wrong. Besides, getting all panicky won't solve anything. Keep your wits about you and proceed logically, and you can probably find what's causing your problem and get it fixed.

2. Check for operator errors. That is, look for something that you did wrong. Maybe you clicked the wrong button, pressed the wrong key, or plugged something into the wrong port. Retrace your steps and try to duplicate your problem. Chances are the problem won't reoccur if you don't make the same mistake twice.

3. Check that everything is plugged in to the proper place and that the system unit itself is getting power. Take special care to ensure that all your cables are securely connected—loose connections can cause all sorts of strange results.

4. Make sure you have the latest versions of all the software and apps installed on your system. That's because old versions of most programs probably haven't been updated with the latest bug fixes and compatibility patches. (These are small updates that typically fix known issues within a program.)

5. Try to isolate the problem by when and how it occurs. Walk through each step of the process to see if you can identify a particular program or process that might be causing the problem.

6. When all else fails, call in professional help. If you have a brand-new PC and you think it's a Windows-related issue, contact Microsoft's technical support department. If you think it's a problem with a particular program or app, contact the tech support department of the program's manufacturer. If you think it's a hardware-related problem, contact the manufacturer of your PC or the dealer you bought it from. The pros are there for a reason—when you need technical support, go get it.

>>>Go Further

GETTING HELP

Many computer users easily become befuddled when it comes to dealing with even relatively simple computer problems. I understand completely; there's little that's readily apparent or intuitive about figuring out how to fix many PC-related issues.

If you feel over your head or out of your element when it comes to dealing with a particular computer problem, that's okay; you don't have to try to fix everything yourself. You have many options available to you, from Best Buy's ubiquitous Geek Squad to any number of local computer repair shops. Google **computer repair** for your location, check your local Yellow Pages, or just ask around to see who your friends use for computer support. It might prove faster and less aggravating in the long run to pay a professional to get your computer working properly again.

← Settings

Michael Miller
trapperjohn2000@hotmail.com

Find a setting

- System
- Bluetooth & devices
- Network & internet
- Personalization
- Apps
- Accounts
- Time & language
- Gaming
- Accessibility
- Privacy & security
- Windows Update

Windows Update

You're up to date
Last checked: Today, 1:06 PM

Check for updates

More options

⏸ Pause updates   Pause for 1 week ⌄

🕐 Update history   >

⚙ **Advanced options**   >
Delivery optimization, optional updates, active hours, other update settings

🗔 **Windows Insider Program**   >
Get preview builds of Windows to share feedback on new features and updates

🔍 Get help

💬 Give feedback

In this chapter, you find out how to manage the Windows Update process—and reset your computer if necessary.

→ Managing Windows Update
→ Resetting Your Computer

23

Updating Windows

Microsoft thinks of Windows as a dynamic thing. It's not just a piece of software you install once and never touch again; it's an operating system and environment that is constantly being tweaked and updated to better meet the needs of its users.

To that end, Windows receives regular updates that are delivered over the Internet. More often than not, these updates fix bugs and security issues, but sometimes they deliver new or changed features. It's important to manage how your computer receives these updates so that you get the patches you need with minimal disruption to your daily routine.

Managing Windows Update

Microsoft delivers two types of updates to Windows: feature updates and quality updates. As the name implies, feature updates add new features to the basic operating system. With Windows 11, these are big updates delivered once a year.

Quality updates are smaller updates delivered once a month, typically on the second Tuesday of each month. (Tech folks call this day Patch Tuesday.) These updates offer quality improvements that address bugs, security issues, and the like.

All of these updates are delivered via the Windows Update service. You can control, to a degree, when these updates are installed on your computer, as well as view which updates have been previously installed.

Need a Connection

Windows updates are delivered automatically over the Internet. If your computer isn't connected to the Internet, you don't get the updates—which is not a good thing.

View Update History

Windows keeps a detailed history of which updates are installed on your computer. This may be useful if you need to know precisely which version of Windows your PC is running.

1. From the Start menu, click or tap Settings to open the Settings app.

2. Click or tap Windows Update in the left column.

3. Click or tap Update History.

4. Recent updates are displayed here, organized by type of upgrade—Feature Updates, Quality Updates, Driver Updates (for hardware drivers), Definition Updates (for Windows Security's antimalware detection), and Other Updates.

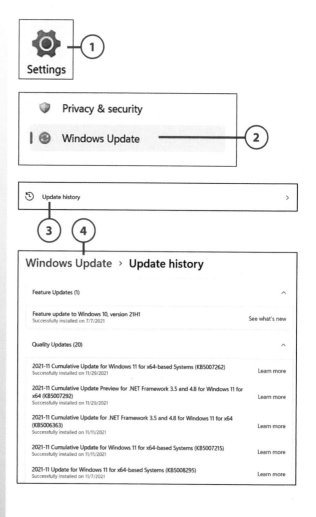

Uninstall an Update

On occasion, Microsoft distributes an update that causes more problems than it solves. A bad update can affect Wi-Fi connectivity, app operation, or system performance, and sometimes it causes Windows to crash. It doesn't happen often, but when an update causes problems (like slowing down or crashing your computer), you may need to uninstall it.

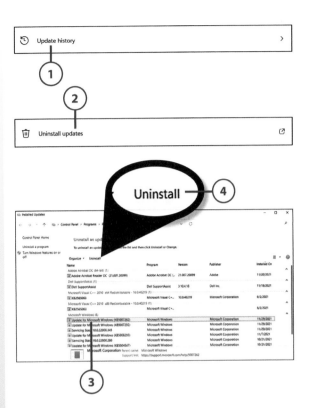

1. Open the Windows Update page in the Settings app and click or tap Update History.

2. Scroll down and click or tap Uninstall Updates to open the Uninstall an Update window.

3. Click or tap the update you want to uninstall.

4. Click or tap Uninstall.

Set Hours Not to Update

Windows updates install automatically but take full control of your computer while they're installing. That means you can't use your computer while updates download and install. Depending on the size of the update and the speed of your Internet connection and computer, the time you can't use your computer can last anywhere from 10 minutes to more than an hour.

For this reason, you probably want to not update Windows during your most active hours, instead scheduling updates for when you're not using your computer—typically in the overnight hours. Windows can do this automatically, based on your typical computer activity.

(1) From the Windows Update page in the Settings app, click or tap Advanced Options.

(2) Click or tap Active Hours to expand this section.

(3) By default, Windows examines your computer usage and automatically sets active hours (during which your computer won't update). To manually set your active hours, click or tap the Adjust Active Hours control and select Manually.

(4) Set the Start Time and End Time for your active hours. Windows will not update during these hours.

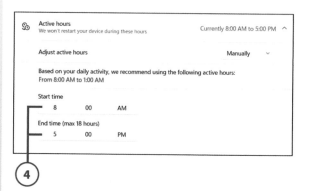

Delay Updates

If you're working on a big project and don't want to risk your computer misbehaving after an update, you may want to delay updates for a bit and not install on Patch Tuesday. Windows 11 lets you pause updates for a specified period of time.

(1) From the Windows Update page in the Settings app, scroll to the Pause Updates section.

(2) Click or tap the control to select how long you want to pause updates, from one to five weeks.

Install an Update Manually

Windows notifies you when an update is available and, in most cases, installs that update automatically during nonactive hours. If you have paused updates, you can opt to install any pending updates manually.

1. Go to the Windows Update page in the Settings app. Any pending updates are listed here.

2. Click or tap Download & Install to download the update now and then install it.

3. If an update has already been updated but not installed, you need to restart the computer to install the update. Click or tap Restart Now to restart your computer and install the update. *Or…*

4. Click or tap Schedule the Restart to schedule the restart for a time when you're not using your computer.

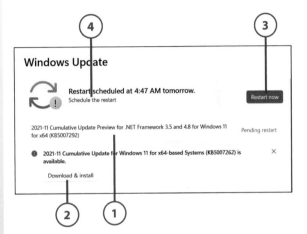

Resetting Your Computer

If your computer frequently freezes or crashes, the problems may be caused by the Windows system files getting damaged or deleted. If this happens, Windows lets you reset your computer's system files with the original versions of these files.

Reset This PC

The Reset This PC tool works by checking whether key system files are working properly. If Windows finds an issue with any files, it attempts to repair those files—and only those files.

This option is somewhat drastic in that it removes all your apps and settings. That means you'll need to reinstall all your apps and reconfigure all your settings after the tool has done its job.

There is also the option of keeping or deleting your personal documents and files. The Keep My Files option deletes apps and settings but keeps your files; the more drastic Remove Everything option deletes your personal files, too.

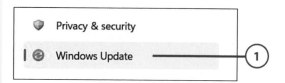

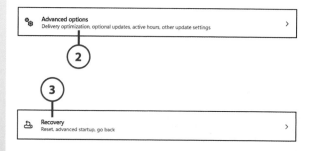

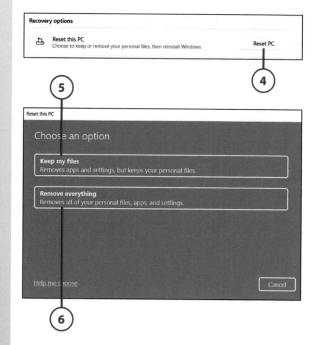

1 From the Settings app, click or tap Windows Update in the left column.

2 Scroll down the right column and click or tap Advanced Options.

3 In the Additional Options section, click or tap Recovery.

4 In the Recovery Options section, click or tap Reset PC.

5 In the Choose an Option window, click or tap Keep My Files if you don't want to delete your personal files and documents. Or...

6 Click or tap Remove Everything to totally wipe everything from your system and return your computer to its original state.

7. When prompted, click Cloud Download to download the latest version of necessary system files. (This downloads the system files from the Internet.) *Or…*

8. Click Local Reinstall to reinstall the original system files stored on your computer's hard disk.

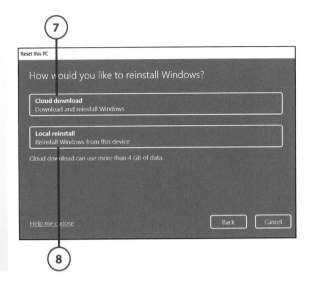

Reset this PC

How would you like to reinstall Windows?

Cloud download
Download and reinstall Windows

Local reinstall
Reinstall Windows from this device

Cloud download can use more than 4 GB of data.

Help me choose   Back   Cancel

It's Not All Good

Everything Is Deleted

The Remove Everything option completely deletes all the files, documents, and programs you have on your system. You'll want to back up your files before taking this extreme step, and then restore your files from the backup and reinstall all the apps you use.

24

Frequently Asked Questions

Do you have a specific question about choosing a new computer, using your computer, or using Windows 11 on your new computer? If that question wasn't answered previously in this book, maybe it's here in this chapter.

Questions About Choosing a New Computer

Do you have questions about choosing the right computer for your particular use? Here are some answers.

How Does a Windows PC Differ from a Mac?

Computers that run Microsoft's Windows operating system differ both subtly and significantly from computers running Apple's MacOS. In the first place, only computers manufactured by Apple can run MacOS; there are numerous manufacturers who make computers that run

Microsoft Windows. Second, MacOS has a completely different look and feel than Windows; the interface is different, basic operations are different—not better and not worse (although the Apple faithful claim otherwise), just different. Third, because Apple is the only company that makes Macs, they tend to be somewhat (okay, significantly) more expensive than Windows PCs.

Can a Mac do things that a Windows PC can't? Not really, although there are some applications that are only available for Macs or that are optimized for use on Macs. In addition, some professions—professional audio recording and mixing, professional video editing, professional photo editing—for whatever reason gravitate toward the MacOS platform. But just about anything you can do on a Mac, you can do on a Windows PC, and vice versa. So if price matters—and it does, for many people—and you want compatibility with 70% or so of computer users out there, Windows is the preferred platform.

How Does a Windows PC Differ from a Chromebook?

Chromebooks are computers that run Google's ChromeOS operating system. ChromeOS is a fast and lightweight operating system that resembles Windows but doesn't run Windows (or Mac) apps. Most Chromebooks are notebooks with smaller (14" or under) screens and very little internal storage. Chromebooks run cloud-based apps rather than desktop apps, so you'll use Google Docs instead of Microsoft Office. They're typically lower-priced than comparable Windows PCs and are very popular in the education market.

If you want to run more powerful desktop apps, you'll probably want to move up to a full-fledged Windows PC. If all you want to do is browse the Web, participate in social media, and watch streaming video, a Chromebook is a less-costly option.

How Much Memory and Storage Space Do I Really Need?

The amount of memory (RAM) you need depends on how you're using your PC. You can run Windows and browse the Web with a bare minimum of 4GB of RAM, but Windows and other apps run a lot faster if your PC has at least 8GB. If you work with large documents or do heavy-duty tasks like video editing or high-end gaming, moving to 12GB or even 16GB would be a good idea.

In terms of internal storage, how much you need depends on how many apps you want to install and what type of—and how many—files you store. For simple web browsing and social networking, you can get by with 256GB storage. If you store a lot of photos, videos, or music, however, you want more storage—512GB is a good number to shoot for. Although you can find PCs with 1TB or more of internal storage, the move to cloud-based storage typically obviates the need for such larger drives.

Should I Buy an Intel or AMD PC?

Two companies make the microprocessor chips inside today's Windows PCs—Intel and AMD. Each company makes a variety of chips at different price points with different processing power. You can find both lower-end and higher-end PCs running both Intel and AMD chips.

Some techies might recommend a particular AMD chip or a specific Intel chip for specific applications, but in reality, both companies make terrific chips that perform similarly. It used to be that some apps might run better on either an Intel or an AMD chip, but that isn't the case anymore. Feel confident buying a PC with either an Intel or an AMD microprocessor, as long as the chips have similar specs.

Do I Need a Touchscreen Display?

Unless you're using your computer as a tablet, as with 2-in-1 models, you don't absolutely, positively need a touchscreen display. From experience, however, I can tell you that touchscreens are nice to have; you'd be surprised how often you'll find yourself tapping the screen even though you have a touchpad and keyboard right in front of you. So while you probably don't *need* a touchscreen PC, you might want one.

What Does Spending More Money Get Me?

Spending more money on a higher-priced PC typically gets you some or more of the following:

- Bigger screen
- Faster processor (so the PC runs faster)

- More memory (RAM), which also makes your PC run faster

- More internal storage for storing more or bigger files

- Faster SSD storage instead of a slower hard drive

If any or all these things matter to you, then give yourself a bigger budget. If all you do is browse the Web and social networks and you don't store a lot of big files, then save your money and buy a lower-priced model.

Questions About Using Your Computer

If you have specific questions about using your new computer or specific apps, here are some answers.

Why Is My Computer Slowing Down/How Can I Speed Up My Computer?

Several things can make your computer slow down over time:

- Installing more apps that use more memory and storage space

- Using more internal storage (storing more or bigger files)

- Accidentally installing computer viruses and spyware

In most instances, it's a simple case of your computer getting older. The more you use it, the more apps and files you install, all of which slow things down. A relatively "clean" new computer will run noticeably faster than an older one that has lots of apps installed and files stored.

You can often speed up your computer, at least a little, by removing unused apps and unnecessary files. Wiping everything from your computer and doing a clean reinstall of Windows should also speed things up, although that's a some-what drastic solution.

If your computer slows down suddenly, the cause could be a virus or other type of malware. Make sure you're running Windows Security or another anti-mal-ware utility to seek out any malicious, system-slowing programs.

How Do I Set Up a Home Network?

Most people use wireless home networks to share broadband Internet connections, printers and other peripherals, and files from one PC to another. A typical home network includes the following:

- Broadband Internet connection, typically coming into a…

- Modem, which translates the digital Internet signals into something all your devices can use and is in turn connected to a…

- Wireless router, which broadcasts the Internet signal to all nearby devices connected via Wi-Fi.

In some instances, the modem and router are combined into a single device called an Internet gateway. (ISPs tend to like this solution.) You can also extend the range of your wireless network by adding wireless extenders or going with so-called mesh networks that include two or more mesh routers to spread the signal across a large house.

Connecting all this together is relatively easy these days; just follow the instructions supplied with the router and you should be good to go. You then connect each of your computers and mobile devices to the network. Windows should recognize other computers connected to the network and let you share files between them, using File Explorer.

How Fast an Internet Connection Do I Need?

The simple answer to this one is, "as fast as you can afford," but it isn't as easy as that. It all depends on what you're doing on the Internet—and how many other people are using it at the same time.

As an example, many people use the Internet to watch streaming video on their smart TVs, streaming media devices, and computers. Netflix recommends a 5Mbps (megabit per second) Internet connection to stream high-definition (HD) content or a 25Mbps connection for 4K video. That's to one device. If you have three people in your household all watching 4K content on their devices at the same time, you'll need a 75Mbps connection to cover that. Add in the bandwidth required to do video chat via Microsoft Teams or Zoom (2Mbps, at least), surf the Web (another 2Mbps or so per device), or play online games (25Mbps or more), and the bandwidth adds up.

If you want to simplify things, follow my recommendations in Table 24.1. There's never any harm in getting more than the recommended minimum, either.

Table 24.1   Recommended Internet Connection Speeds

Number of Devices	Typical Usage	Recommended Minimum Internet Download Speed
1 to 2	Web browsing, social networking, email, video chat, video streaming (HD)	25Mbps
3 to 4	4K video streaming, online gaming	50 to 100Mbps
5+	All the above plus uploading/downloading large files	200+ Mbps

What Do the Function Keys Do?

Function keys are those keys typically lined up across the top of a computer keyboard, labeled F1, F2, F3, and so forth. They're typically used as shortcut keys for specific functions either in Windows or in specific apps. For example, F1 is the help key in many apps, whereas F2 is the shortcut for renaming a file or folder in File Explorer.

My Computer's Running Out of Storage Space—What Do I Do?

There are three possible solutions when your storage space (either hard disk or SSD) is running low. You can either

- Delete unused files and apps on your computer to free up some space

- Invest in more storage, in the form of an external hard drive or SSD

- Move some of your files to Microsoft OneDrive—and begin storing all your new files in the cloud

How Do I Clean My Computer?

Computers are relatively easy to care for, especially if you keep it away from dirt and dust and don't spill anything on it. Do the following:

- Use a small vacuum cleaner to periodically sweep the dirt from your keyboard or use compressed air to blow the dirt away.

- Use a cotton swab or soft cloth to clean between the keys.

- To clean your display and touchpad, dampen a lint-free cloth with water and then use that to wipe the screen. You can also use commercial cleaning wipes or sprays specially formatted for LCD screens, but don't use any cleaner that contains alcohol or ammonia because these chemicals can damage the screen.

- If your mouse has a rollerball underneath, use compressed air to blow dust and dirt from around the bottom.

I Just Spilled Liquid on My Keyboard—What Do I Do?

If you spill something on your keyboard, either disconnect the keyboard (if it's a desktop or all-in-one PC) or turn off the entire PC if you have a laptop. Use a soft cloth to get between the keys; if necessary, use a screwdriver to pop off the keycaps and wipe up any seepage underneath. Let the keyboard dry thoroughly before trying to use it again.

If liquid seeps beneath the keyboard on a notebook PC, it can damage the electronics underneath. This is not a good thing and, if you notice your computer acting erratically after a spill, might require you to take it to a technician for further evaluation.

Questions About Using Windows

Questions about Windows? You're not alone; here are answers to some of the most common questions that users have about Microsoft's operating system.

What Is Safe Mode and How Do I Get into It?

Safe mode is a special running mode that runs a barebones version of Windows, without a lot of drivers and add-ins. It's useful for diagnosing serious operating system problems.

To enter Safe mode, reboot your computer and, as the computer is booting up, press and hold the F8 key on your keyboard. This should interrupt the normal boot process and display the Windows Advanced Options menu. Use your keyboard's arrow keys to select Safe Mode from the available options and then press Enter. When Windows restarts, it should be in Safe mode.

When I Try to Delete a File, I Get an Error Message That It's Being Used By Another Program—What Do I Do?

You can't delete files that are open in other programs. If you can identify which program that file is open in, close it and then try deleting the file again. If you can't easily identify which program is using the file, close all your open files and then try deleting the file. If that still doesn't work, reboot your computer and delete the file then.

Is It Safe to Turn Off My Notebook PC Without Doing a Windows Shut Down?

If you have a notebook or 2-in-1 PC, you may think you can shut down your PC by closing the lid or pressing the Power button. These approaches don't always shut down your computer, however; more often than not, they just put your computer into sleep mode.

To fully exit Windows and properly shut down your computer, you should open the Windows Start menu, click or tap the Power icon, and then select Shut Down. This official shut-down process goes through all the necessary actions to close all open programs and exit Windows properly.

I Accidentally Deleted a File—How Can I Get It Back?

Deleting a file in Windows doesn't actually delete it; it just sends the file to the Recycle Bin. "Deleted" files are stored in the Recycle Bin for a specified period or until that disk space is needed for new files and programs, which means you can recover any files you accidentally delete. Just open the Recycle Bin, select the file you want to recover, and then click or tap Restore the Selected Item(s).

Questions About Safe Computing

Are you worried about computer viruses and cyberattacks? Here are some questions, answered.

Protecting Yourself Online

Learn more about Windows Security and safe computing in Chapter 13, "Protecting Yourself Online."

Do I Need to Buy an Antivirus Program?

If you just purchased a new PC, it probably came with a preinstalled trial version of an antivirus program. (And that trial version is probably constantly nagging you to upgrade to the full version.) However, purchasing that or any other antivirus program probably isn't necessary because Windows 11 comes with a pretty good antivirus program, called Windows Security, built in. There is little in the way of additional anti-malware protection a third-party program may offer—although many third-party programs include additional functionality that you might deem useful.

How Do I Know If My Computer Has a Virus?

If you notice any of the following issues with your computer, it may be infected with a virus or other malicious software:

- PC suddenly runs slower than normal.
- PC freezes.
- PC shuts down or restarts all by itself.
- Individual programs crash.

- You see an unexpected error message on the screen.
- Files are missing.
- Your computer's hard drive makes odd sounds or seems to be working harder than normal, especially if you're not doing much on the PC.
- Your email program is sending out emails all by itself.
- Web browser is slow to load common websites.
- Web browser automatically redirects to strange websites.

This is especially true if you notice any of these symptoms after opening an email file attachment, downloading a file from the Internet, or installing a new app.

What Do I Do If My Computer Is Infected with a Virus?

If you think your computer is infected with a virus or other malware, run Windows Security or another anti-malware utility. If the virus doesn't let you run the anti-malware tool, then you may need to take your PC into a technician for further evaluation.

I Got an Email with a File Attached—Should I Open It?

No—unless you specifically expected the attachment from a known source. The majority of malware is spread via files attached to email messages.

I Got an Email Asking Me for Private Information—What Should I Do?

Don't provide it. Don't click any links in the email. If you think it's a known source—like your bank—requesting the information, use your web browser to access the source's site manually. You'll probably find that there are no messages waiting for you and no action required on your part—which means the email was a phishing scheme trying to scam you out of your personal information.

Glossary

1–10

2-in-1 computer   A portable computer that combines the functionality of a touchscreen tablet and traditional notebook PC.

A

address   The location of an Internet host. An email address might take the form johndoe@xyz.com; a web address might look like www.xyztech.com. See also *URL*.

all-in-one computer   A desktop computer where the system unit, monitor, and speakers are housed in a single unit. Often the monitor of such a system has a touchscreen display.

app   *See application.*

application   A computer program designed for a specific task or use, such as word processing, accounting, or missile guidance.

attachment   A file, such as a Word document or graphics image, attached to an email message.

B

backup   A copy of important data files.

boot   The process of turning on your computer system.

broadband   A high-speed Internet connection; it's faster than the older dial-up connection.

browser   A program, such as Microsoft Edge or Google Chrome, used to view pages on the Web.

bug   An error in a software program or the hardware.

byte   A group of 8 binary digits (bits), used as a measurement of memory or other storage.

C

computer   A programmable device that can store, retrieve, and process data.

CPU (central processing unit)   The group of circuits that direct the entire computer system by (1) interpreting and executing program instruction and (2) coordinating the interaction of input, output, and storage devices.

Craigslist   An Internet-based classified advertising forum.

cursor   The highlighted area or pointer that tracks with the movement of your mouse or arrow keys onscreen.

D

data   Information—on a computer, in digital format.

desktop   The background in Windows upon which all other apps and utilities sit.

desktop computer   A personal computer designed for use on a typical office desktop. A traditional desktop computer system consists of a system unit, monitor, keyboard, mouse, and speakers.

device   A computer file that represents some object—physical or nonphysical—installed on your system.

domain   The identifying portion of an Internet address. In email addresses, the domain name follows the @ sign; in website addresses, the domain name follows the www.

download   A way to transfer files, graphics, or other information from the Internet to your computer.

driver   A support file that tells a program how to interact with a specific hardware device, such as a hard disk controller or video display card.

E

email   Electronic mail; a means of corresponding with other computer users over the Internet through digital messages.

encryption   A method of encoding files so only the recipient can read the information.

Ethernet   A popular computer networking technology; Ethernet is used to network, or hook together, computers so that they can share information.

executable file   A program you run on your computer system.

F

favorite   A bookmarked site in a web browser.

file   Any group of data treated as a single entity by the computer, such as a word processor document, a program, or a database.

File Explorer   The utility used to navigate and display files and folders on your computer system.

firewall   Computer hardware or software with special security features to safeguard a computer connected to a network or to the Internet.

folder   A way to group files on a disk; each folder can contain multiple files or other folders (called *subfolders*). Folders are sometimes called *directories*.

freeware   Free software available over the Internet. This is in contrast with *shareware*, which is available freely but usually asks the user to send payment for using the software.

G

gigabyte (GB)   One billion bytes.

graphics   Pictures, photographs, and clip art.

H

hard disk drive   A mechanism for long-term storage of digital data on a spinning magnetic disk. The drive rotates the disk at high speed and reads the data with a magnetic head.

hardware   The physical equipment, as opposed to the programs and procedures, used in computing.

HDMI (high-definition multimedia interface)   An interface for transmitting high-definition digital audio and video signals.

home page   The first or main page of a website.

hotspot   A public wireless Internet access point.

hover   *See mouse over.*

hyperlink   A connection between two tagged elements in a web page, or separate sites, that makes it possible to click from one to the other. Sometimes called a *web link*.

I–J

icon   A graphic symbol on the display screen that represents a file, peripheral, or some other object or function.

identity theft   The illegal use of a person's personal information, typically for fraudulent purposes.

Internet   The global network of networks that connects millions of computers and other devices around the world.

Internet service provider (ISP)   A company that provides end-user access to the Internet via its central computers and local access lines.

K–L

keyboard   The typewriter-like device used to type instructions to a personal computer.

kilobyte (KB)   A unit of measure for data storage or transmission equivalent to 1024 bytes; often rounded to 1000.

LAN (local-area network)   A system that enables users to connect PCs to one another or to minicomputers or mainframes.

laptop   A portable computer small enough to operate on one's lap. Also known as a *notebook* computer.

M–N

malware   Short for *malicious software*, any software program designed to do damage to or take over your computer system.

megabyte (MB)   One million bytes.

megahertz (MHz)   A measure of microprocessing speed; 1MHz equals one million electrical cycles per second. (One thousand MHz equals 1 gigahertz, or GHz.)

memory   Temporary electronic storage for data and instructions, via electronic impulses on a chip.

microprocessor   A complete central processing unit assembled on a single silicon chip.

Microsoft Edge   The web browser included with Windows 11.

Microsoft Store   Microsoft's online store that offers Windows apps for sale and download.

modem (modulator demodulator)   A device capable of converting a digital signal into an analog signal, typically used to connect to the Internet.

monitor   The display device on a computer, similar to a television screen.

motherboard   Typically the largest printed circuit board in a computer, housing the CPU chip and controlling circuitry.

mouse   A small handheld input device connected to a computer and featuring one or more button-style switches. When moved around on a flat surface, the mouse causes a symbol on the computer screen to make corresponding movements.

mouse over   The act of selecting an item by placing your cursor over an icon without clicking. Also known as *hovering*.

network   An interconnected group of computers.

notebook computer   A portable computer with all components (including keyboard, screen, and touchpad) contained in a single unit. Notebook PCs can typically be operated via either battery or wall power. Also known as a *laptop*.

Notifications panel   A panel that displays system notifications and messages from individual apps. Displays when the right side of the notifications area of the taskbar is clicked.

O–P

operating system   A sequence of programming codes that instructs a computer about its various parts and peripherals and how to operate them. Operating systems, such as Windows, deal only with the workings of the hardware and are separate from software programs.

path   The collection of folders and subfolders (listed in order of hierarchy) that hold a particular file.

peripheral   A device connected to the computer that provides communication or auxiliary functions, such as a keyboard, mouse, or printer.

phishing   The act of trying to "fish" for personal information via means of a deliberately deceptive email or website.

pixel   The individual picture elements that combine to create a video image.

port   An interface on a computer to which you can connect a device, either internally or externally.

printer   The piece of computer hardware that creates hard copy printouts of documents.

Q–R

Quick Settings panel   A panel that provides access to frequently used Windows controls. Displays when the left side of the notification area on the taskbar is clicked.

RAM (random-access memory)   A temporary storage space in which data can be held on a chip rather than being stored on disk or tape. The contents of RAM can be accessed or altered at any time during a session but will be lost when the computer is turned off.

resolution   The degree of clarity an image displays, typically expressed by the number of horizontal and vertical pixels or the number of dots per inch (dpi).

ribbon   A toolbar-like collection of action buttons, used in many Windows programs.

ROM (read-only memory)   A type of chip memory, the contents of which have been permanently recorded in a computer by the manufacturer and cannot be altered by the user.

root   The main directory or folder on a disk.

router   A piece of hardware or software that handles the connection between your home network and the Internet.

S

scanner   A device that converts paper documents or photos into a format that can be viewed on a computer and manipulated by the user.

server   The central computer in a network, providing a service or data access to client computers on the network.

shareware   A software program distributed on the honor system; providers make their programs freely accessible over the Internet, with the understanding that those who use them will send payment to the provider after using them. See also *freeware*.

Snap Layouts   Easy-to-use onscreen layouts for use with multiple open windows.

software   The programs and procedures, as opposed to the physical equipment, used in computing.

solid-state drive (SSD)   A form of long-term data storage that stores data electronically with no moving parts.

spam   Junk email. As a verb, it means to send thousands of copies of a junk email message.

spreadsheet   A program that performs mathematical operations on numbers arranged in large arrays; used mainly for accounting and other record keeping.

spyware   Software used to surreptitiously monitor computer use (that is, spy on other users).

Start menu   The pop-up menu, activated by clicking the Start button, that displays all installed programs on a computer.

system unit   The part of a desktop computer system that looks like a big gray or black box. The system unit typically contains the microprocessor, system memory, hard disk drive, floppy disk drives, and various cards.

T–U–V

tablet computer   A small, handheld computer with no keyboard or mouse, operated solely via its touchscreen display.

Task View   The Windows function that enables the creation of multiple virtual desktops, each with its own collection of open apps.

taskbar   The bar at the bottom of the Windows screen that allows quick access to apps and functions.

terabyte (TB)   One trillion bytes.

touchpad   The pointing device used on most laptop PCs, in lieu of an external mouse.

touchscreen display   A computer display that is touch sensitive and can be operated with a touch of the finger.

Trusted Platform Module (TPM)   A special computer chip that uses a cryptographic key to secure computer hardware. Windows 11 only runs on computers with TPM chips version 2.0 and above.

upgrade   To add a new or improved peripheral or part to your system hardware. Also to install a newer version of an existing piece of software.

upload   The act of copying a file from a personal computer to a website or Internet server. The opposite of *download*.

URL (uniform resource locator)   The address that identifies a web page to a browser. Also known as a *web address*.

USB (universal serial bus)   The most common type of port for connecting peripherals to personal computers.

virus   A computer program segment or string of code that can attach itself to another program or file, reproduce itself, and spread from one computer to another. Viruses can destroy or change data and in other ways sabotage computer systems.

W–X–Y–Z

web page   An HTML file, containing text, graphics, and/or mini-applications, viewed with a web browser.

website   An organized, linked collection of web pages stored on an Internet server and read using a web browser. The opening page of a site is called a *home page*.

Wi-Fi   The radio frequency (RF)–based technology used for home and small business wireless networks and for most public wireless Internet connections. Short for wireless fidelity.

widget   A small panel that displays specific information. In Windows 11, widgets are displayed in the Widgets panel that pulls in from the left side of the screen.

window   A portion of the screen display used to view simultaneously a different part of the file in use or a part of a different file than the one in use.

Windows   The generic name for all versions of Microsoft's graphical operating system.

Windows Security   The suite of tools built into Windows 11 to protect against malware, computer attacks, and other unwanted intrusions.

Windows Update   The service built into Windows that automatically manages updates to the operating system.

World Wide Web (WWW)   A vast network of information, particularly business, commercial, and government resources, that uses a hypertext system for quickly transmitting graphics, sound, and video over the Internet.

Index

C

F

H

I

J–K

L

M

N

O

P

X–Y

Z

Answers to Your Technology Questions

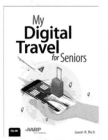

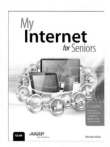

The **My...For Seniors Series** is a collection of how-to guide books from AARP and Que that respect your smarts without assuming you are a techie.
Each book in the series features:

- Large, full-color photos
- Step-by-step instructions
- Helpful tips and tricks
- Troubleshooting help

For more information about these titles,
and for more specialized titles, visit
informit.com/que

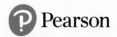